Post Mortem Changes in Fish

Post Mortem Changes in Fish

Dr. B. K. Khuntia
Associate Professor,
College of Fisheries (OUAT),
Rangailunda, Berhampure - 760 007
Orissa

2009
DAYA PUBLISHING HOUSE
Delhi - 110 035

ISBN 81-7035-602-4
ISBN 978-81-7035-602-8

Published by : **Daya Publishing House**
1123/74, Deva Ram Park
Tri Nagar, Delhi - 110 035
Phone: 27383999
Fax: (011) 23260116
e-mail : dayabooks@vsnl.com
website : www.dayabooks.com

Showroom : 4760-61/23, Ansari Road, Darya Ganj,
New Delhi - 110 002
Phone: 23245578, 23244987

Laser Typesetting : **Classic Computer Services**
Delhi - 110 035

Printed at : **Chawla Offset Printers**
Delhi - 110 052

PRINTED IN INDIA

To

My Wife
SASMITA

Acknowledgement

I express my heartfelt thanks to Dr. P.C. Thomas, Ex-Director, College of Fisheries (OUAT), Rangailunda, for his constant encouragement, critical comments and constructive suggestions about this book, which inspired me to get it published. I am extremely thankful to my colleagues at College of Fisheries (OUAT), Rangailunda who have extended their kind cooperation during writing of this book. Their critical suggestions have enabled me to upgrade the quality of the book.

I am indebted to my father Sj. Jayakrishna Khuntia and mother Smt. Taapoi Khuntia whose generous blessings have enabled me to write this book. I owe a lot to my wife Sasmita, for her constant encouragement and deliberate support in bringing the book to the present form. I am thankful to my daughters Nilima and Banani for their patience and cooperation during writing of the book.

I am greatly obliged to Dr. S. Ayyappan, Deputy Director General (Fisheries), Indian Council of Agricultural Research for kindly writing a foreword to the book. I also extend my thanks to Daya Publishing House, New Delhi for publishing the book in such an excellent manner.

B. K. Khuntia

Foreword

The demand for fish is increasing day-by-day due to the rapid growth of population and steep rise in the consumers' demand for healthy foods. At present, fish is not only treated as a cheap and easily available source of animal protein, it has also acquired an esteemed position as a rich source of ω-3 fatty acids, the panacea for cardiovascular diseases. It has surpassed many food items to become one of the most popular healthy foods all over the world. At the same time the consumer is becoming more and more conscious about the quality of fish and fishery products available in the market. On the other hand, the production of fish has reached an almost stagnation in recent years. This invites for the effective utilisation of the produce without allowing a bit of it to be wasted. As fish is a highly perishable food commodity, it needs greater care than most of the other food materials.

Several changes occur in the fish body after death. These post mortem changes gradually lead to the loss of quality ultimately resulting in its spoilage. A thorough understanding of these post mortem changes is required to meticulously manipulate them so as to retain its quality over a prolonged period of time. This will not only help in providing the consumer with good quality fish and fishery products, it will also reduce the wastage of fish resulting from its spoilage. This book entitled "Post Mortem Changes in Fish" authored by Dr. B. K. Khuntia, Associate Professor, College of Fisheries, Orissa University of Agriculture and Technology, contains an in-depth description of these post mortem changes such as hyperaemia, rigor mortis, autolysis, microbial putrefaction, lipid autoxidation and discolouration. The intricate mechanisms of these changes have been described in a lucid manner with several illustrations.

I congratulate Dr. Khuntia for his untiring efforts in preparing such an exhaustive and illustrated book. I am sure this book will be an invaluable asset for the students, teachers, researchers, professionals, traders, fish processing technologists and all those involved in fish processing.

Dr. S. Ayyappan
Deputy Director General (Fisheries)
Indian Council of Agricultural Research
New Delhi

Preface

Fish, as a food commodity, has been associated with man since time immemorial. It is a nutritionally rich, cheap source of animal protein. At present, it is gaining additional importance among the affluent people due to their increasing awareness about its unique nutritional properties not found in any other food commodity. Although advancement in technology of fishing and aquaculture has resulted in increasing the production of fish, its effective utilisation has not yet met with the expected success, particularly in the tropical developing countries. A large portion of the total world fish catch is either discarded or sold at a low price due to the fast deterioration of its quality. The problem is more precarious in the tropical countries where the high ambient temperature further enhances the rate of quality deterioration.

Fish is a highly perishable food commodity as its quality starts deteriorating immediately after death. The lion's share of the total world fish catch is consumed fresh in which the post mortem changes have tremendous influence on quality. Moreover, the quality of processed fishery products depends largely on the quality of the raw material which in turn depends on the post mortem changes. As no processing method can improve the quality of fish, it is imperative to take utmost care to minimise the post mortem deteriorative processes occurring in the fish body. This necessitates a thorough understanding of the processes, which occur in the fish body after death so that the factors influencing these processes can be manipulated to minimise the deteriorative processes. The book has been written to provide some basic knowledge about the post mortem changes which occur in fish and the ways they can be controlled desirably.

The book elaborates various aspects of the post mortem changes in fish through a number of illustrations. There are seven Chapters in the book, each followed by a summary. In addition, there are nine Text Boxes, which provide further information related to the post mortem changes. The book may be introduced as a textbook at postgraduate level for students specialising in fish processing. For undergraduate students of fisheries, the summary at the end of each chapter is quite adequate to brief them the fundamentals of changes in fish body after death. This will open the window for further understanding of other subjects in fish processing. The book shall be useful to students, teachers, researchers, extension personnel and all associated with fish processing. The author cordially invites critical comments and suggestions for further improvement in the quality of the book in future.

B. K. Khuntia

E-mail: basantak@yahoo.com

Contents

List of Figures

List of Tables

List of Text Boxes

1

Introduction

I. What are Post Mortem Changes ?

Fish has been a lucrative food item since ancient times. Besides its use as food, it has gained additional importance in recent years because of its agricultural, industrial, medicinal and ornamental uses. In most cases, its ultimate utilisation needs its 'carcass' or the dead body. For its utilisation, fish is harvested from its aquatic habitat, which leads to its death. After death several changes take place in its body. These changes are collectively called 'post mortem changes' (*post*: after; *mortem*: death). They can be grouped into the following six steps mentioned in the order of their occurrence after death.

1. Hyperaemia

In this step, the skin of fish releases large quantity of mucus to the body surface.

2. Rigor Mortis

During rigor mortis (*rigor*: stiffening; *mortis*: death) the body of fish stiffens for a certain length of time after death.

3. Autolysis

In autolysis (*auto*: self; *lysis*: break down) the complex tissue components of fish body such as proteins, fats (lipids) and nucleic acids are hydrolysed or broken down into their simple building blocks by the enzymes present in the fish body (endogenous enzymes). Proteins are hydrolysed to amino acids; fats to fatty acids and glycerol and nucleic acids to nucleotides as given below. Due to hydrolysis of proteins, which are the important structural component of fish tissue, autolysis results in the softening of fish tissue. The end products of hydrolysis become a nutrient-rich medium for the growth of microbes.

Proteins ⟶ Amino acids

Fats ⟶ Fatty acids + glycerol

Nucleic acids (RNA & DNA) ⟶ Nucleotides

4. Microbial Putrefaction or Decomposition

In this step, the tissue components of fish, intact or hydrolysed through autolysis, are decomposed by microorganisms into off-odour, off-flavour substances accumulation of which distracts the consumers and thereby results in the spoilage of fish.

5. Lipid Oxidation or Autoxidation

In autoxidation, the unsaturated lipids of fish tissue are oxidised by the atmospheric oxygen at their points of unsaturation resulting in the development of rancid flavour in the fish tissue. It is prominent particularly in fatty (oily) fishes whose tissue is rich in lipids, mostly unsaturated types.

6. Discolouration

It is the undesirable changes in the colour of fish tissue after death. The important types of discolouration are the brown discolouration of red-meat fishes like tuna in which the red colour of meat changes to brown and the black discolouration of shrimps in which black spot formation takes place due to improper handling after harvesting.

Although the post mortem changes have been arranged as above in the order of their occurrence, some of them may continue simultaneously, may be at different rates and intensities and some continue sequentially. For example, autolysis and decomposition occur almost simultaneously whereas autolysis follows rigor mortis, thus occurring sequentially. Although autolysis and decomposition begin and continue almost simultaneously, their speed and intensity depend on the storage temperature. At lower temperature, say in ice, the microbial activity is retarded and autolysis becomes more intense than microbial decomposition. Conversely, at higher temperature, microbial decomposition predominates. The post mortem changes have been described in details in the subsequent chapters.

II. Summary

The changes, which take place in the body of fish after death are collectively called post mortem changes (*post*: after; *mortem*: death). They can be grouped into the following six steps.

1. Hyperaemia

The skin releases large quantity of mucus to the body surface.

2. Rigor Mortis (*Rigor*: stiffening; *mortis*: death)

The body of fish stiffens for a certain length of time after death.

3. Autolysis (*Auto*: self; *lysis*: break down)

The complex tissue components of fish body such as proteins, fats (lipids) and nucleic acids are hydrolysed by endogenous enzymes into their simple building blocks. Proteins are hydrolysed to amino acids, fats to fatty acids and glycerol and nucleic acids to nucleotides. Autolysis results in the softening of fish tissues.

4. Microbial Putrefaction or Decomposition

The components of fish tissue are decomposed by microorganisms into off-odour, off-flavour substances, accumulation of which results in the spoilage of fish.

5. Lipid Oxidation or Autoxidation

The unsaturated lipids of fish tissue are oxidised by the atmospheric oxygen resulting in the development of rancid flavour in the fish tissue.

6. Discolouration

It is the undesirable changes in the colour of fish tissue after death.

2

Hyperaemia

I. What is Hyperaemia ?

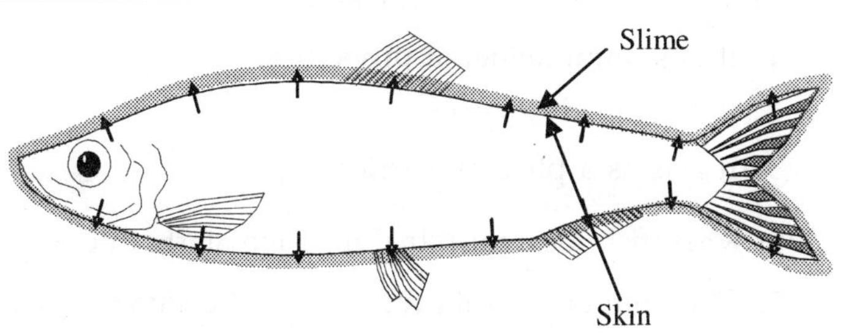

Figure 1: Secretion of slime to body surface during hyperaemia

During harvesting of fish from its aquatic habitat, its immediate surrounding becomes unfavorable to it. Due to this, very often, the fish dies during harvesting. For example, fish caught in a trawl net dies while still inside it due to overcrowding and insufficient oxygen (asphyxia). As a peculiar reaction of the dying fish to the unfavorable surrounding, the mucus glands of the skin secret large quantity of mucus to the skin surface which forms a thick layer of slime on its body (Figure 1). The secretion of slime continues even after death. This phenomenon is called 'hyperaemia'. The secretion of mucus is sometimes so considerable that, the whole body becomes covered by a thick layer of slime accounting to about 2–2.5 per cent of its body weight.

When fish is alive, a thin layer of slime covers its body. It protects the fish body against microbial invasion in three ways as given below.

1. In some fishes the slime contains antibodies, which have immunological properties against microbial invasion into the fish body.
2. In some fishes the slime contains the antimicrobial substance 'lysozyme' that can kill bacteria.
3. The slime layer acts as a physical barrier against microbial attack.

After the death of fish, however, the following events occur.

1. The immunological properties of the slime are lost and production of fresh antibodies (immunoglobins) stops.
2. Microbes attach to the sticky slime and do not get washed away easily.
3. Slime consists mainly of the glycoprotein, mucin, which is a rich nutrient for microbes. It becomes an excellent substrate for microbial growth. The natural microbial flora of the skin grows in the slime rapidly after death.

Due to the above events, the microbes grow profusely in the slime and putrefy it. As the slime putrefies, it gives off an offensive smell. Besides, the profuse growth of the skin microbes increases the chance of their penetration into the sterile muscle and internal organs. Microbes, particularly bacteria, are present most abundantly in the gut of fish, followed by in the gills and on the skin whereas the muscle and internal organs are sterile (free of microbes). The microbes, which grow on the slime, penetrate into the sterile muscle and internal organs through the skin when it becomes permeable to them upon its hydrolysis by the body enzymes during autolysis. Their entry becomes still easier if the fish gets cuts or bruises during handling through which they can enter into the muscle and internal organs. These microbes grow there and decompose them with production of off-odour, off-flavour substances. The presence of these substances distracts the consumers and thereby renders the fish spoiled. Thus the release of slime does not in itself lead to spoilage; rather it acts as a substrate for microbial growth and as a vehicle for penetration of microbes into fish body, which results in its spoilage.

II. Summary

The secretion of large quantity of slime to the body surface after death of fish is called 'hyperaemia'. It occurs as a peculiar reaction of the dying fish to the unfavorable surrounding. The mucus glands of the skin secret large quantity of mucus to the skin surface which forms a thick layer of slime on the fish body. When fish is alive, a thin layer of slime covers its body. It protects the fish body against microbial invasion in three ways as given below.

1. It has immunological properties.
2. It contains lysozyme.
3. It acts as a physical barrier.

But when fish dies, microbial invasion starts due to the following three reasons.

1. The immunological properties of the slime are lost.
2. Microbes attach to the sticky slime.
3. As slime consists mainly of the glycoprotein, mucin, which is a rich nutrient for microbes, it becomes an excellent substrate for growth of the natural microbial flora present on the skin.

These microbes putrefy the slime giving offensive smell. They penetrate into the muscle and internal organs through the skin, which becomes still easier if there are cuts or bruises on it. They decompose the muscle and internal organs with production of off-odour, off-flavour substances, presence of which distracts the consumers and renders the fish spoiled. Thus the release of slime does not in itself lead to spoilage; rather it acts as a substrate for microbial growth and as a vehicle for penetration of microbes into fish body, which results in its spoilage.

3

Rigor Mortis

I. What is Rigor Mortis ?

After a short lapse of time from the death of fish, its body remains stiffened for a certain period during which the elastic nature of living muscle disappears and the muscle becomes hard. Such stiffening of muscle after death is called 'rigor mortis' (*rigor*: stiffening; *mortis*: death). It starts 1-7 hours after death; reaches peak between 5-22 hours and its total duration is 31-120 hours. Though some deteriorative changes take place during rigor mortis, their degree is much less as compared to those taking place during subsequent autolysis and microbial putrefaction. The main factors responsible for spoilage of fish are autolysis and microbial putrefaction. Prolongation of rigor mortis delays the onset of subsequent deteriorative autolysis and microbial putrefaction thereby delaying the onset of spoilage and retaining the freshness of fish for a longer duration. As a result, the shelf life of fish becomes longer. Therefore, all processing steps undertaken immediately after death of fish aim at the prolongation of the duration of rigor mortis. Thus, prolongation of rigor mortis is of economic importance from processing point of view.

Depending upon the state of rigor mortis, fish to be utilised for processing can be in any one of the following three states.

1. Pre-rigor

It is the state of fish carcass from death till the onset of rigor mortis. Fish in this state is extremely fresh. The muscle remains elastic. The time lag between death and onset of rigor mortis is called 'lag of initiation'.

2. On-rigor

It is the state of fish carcass from the onset of rigor mortis till its disappearance. In this state, the fish body remains stiffened as rigor mortis continues.

3. Post-rigor

This state sets in after the disappearance of rigor mortis. In this state the muscle of fish carcass becomes soft and flabby.

When fish is frozen without completing the rigor mortis (pre-rigor or on-rigor), the freezing process arrests the rigor process and the rigor is completed during thawing[1]. This rigor is called 'thaw rigor'. When fish is killed by electric shock, rigor sets in long time after death. Such a rigor is called 'cataleptic rigor'[2].

---

1 Thawing: It is the process of melting hard solidified frozen fish to its normal soft condition before its use.

2 Catalepsy : Disease in which the patient becomes unconscious and his muscles stiffen.

II. Stages of Rigor Mortis

When a fish is alive, its muscle is firm, elastic and has a definite degree of softness, which changes after death. The degree of softness of fish muscle in live condition is called 'original softness level' (Figure 2). It is maintained for a certain lapse of time from death. This period is called pre-rigor period or lag of initiation. After this, rigor mortis sets in manifesting itself in stiffening of the muscle. Rigor mortis passes through two stages as given below.

1. Stiffening of Muscle

During the stiffening stage, the muscle stiffens gradually resulting in a gradual decrease in the softness from the original softness level. Stiffening continues to a point of maximum stiffness (or lowest softness) called 'peak'. After this starts the second stage of rigor mortis, the softening of muscle.

2. Softening of Muscle

During this stage, the muscle loses its stiffness gradually and softens resulting in a gradual increase in its softness. Softening continues till the softness returns back to original softness level. Rigor mortis is said to be completely resolved or terminated when the softness of fish tissue reaches the original softness level. However, the gradual softening of the muscle continues unabated even after the termination of rigor mortis. This results in a gradual increase in the softness of muscle further above the original softness level. Such softening of muscle above the original softness level is the main observable characteristics of the next post mortem stage, autolysis, which follows rigor mortis.

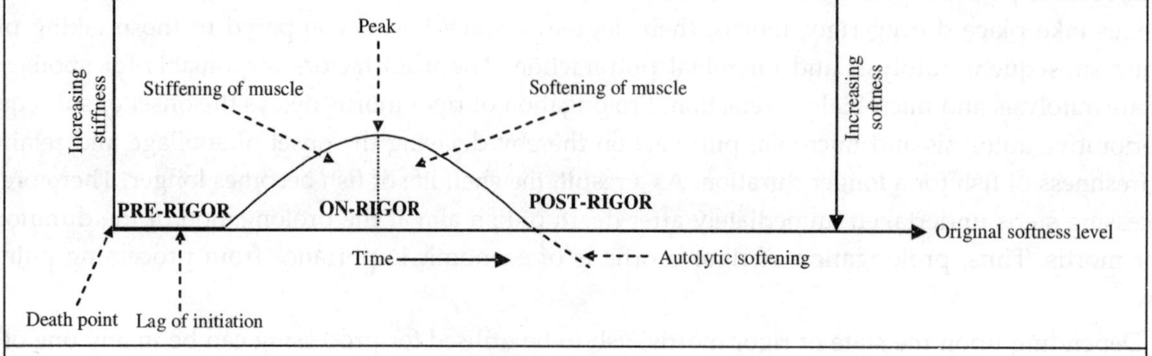

Figure 2: Stages of rigor mortis

III. Biochemistry of Rigor Mortis

After the death of fish a series of events occur in its body. The most important among them, which is responsible for rigor mortis, is the break down of tissue components starting immediately after death.

1. Break Down of Tissue Components

Fish tissue is a complex system consisting of a large number of components. Five of them, which are most important with respect to post mortem changes are carbohydrates, nucleotides, proteins, lipids (fats) and nucleic acids (DNA and RNA). After death of fish, its endogenous tissue enzymes hydrolyse these components. The five post mortem hydrolytic processes are given below.

A	**Saccharolysis**	**Break down of carbohydrates (glucose and glycogen)**	**Start immediately after death**
B	**Nucleotidolysis**	**Break down of nucleotides (ATP)**	
C	Proteolysis	Break down of proteins	Start with onset of autolysis which follows rigor mortis
D	Lipolysis	Break down of lipids (fats)	
E	Nucleolysis	Break down of nucleic acids (DNA and RNA)	

The break down of the carbohydrates (glycogen and glucose) and of the nucleotide (ATP) start immediately after death leading to rigor mortis, while the break down of proteins, lipids and nucleic acids occurs during autolysis which sets in after rigor mortis is terminated. Therefore, saccharolysis and nucleotidolysis are described below while proteolysis, lipolysis and nucleolysis are described under autolysis in the next chapter.

A. Saccharolysis

Saccharolysis (*saccharo*: sugar; *lysis*: break down) is the process of breaking down the carbohydrates or sugar substances by a group of carbohydrate-breaking or saccharolytic enzymes, collectively called 'carbohydrases'. The most important carbohydrates present in fish body are glycogen and glucose. When fish is alive, the glucose absorbed into the body after digestion of the carbohydrates in its food, is polymerised to glycogen by the process called 'glycogenesis'. This glycogen is stored as glycogen granules in the liver cells and muscle cells. Some amount of free glucose (0.1–3. 9 μmol/g) is also available in the muscle of living fish. The carbohydrates in the body are used up by saccharolytic processes (Figure 3). The two most important saccharolytic processes taking place in fish body after death are glycogenolysis and glycolysis.

(*a*) Glycogenolysis

In glycogenolysis, glycogen is hydrolysed to glucose. It can occur in two paths as follows.

(*i*) *Amylolytic Path*

In this path, glycogen is directly hydrolysed to glucose by the enzyme α-amylase. The glucose produced by glycogenolysis is phosphorylated (*i.e.* phosphate is added) to produce glucose-6-phosphate, which goes for anaerobic glycolysis resulting in the production of lactic acid.

(*ii*) *Glycolytic Path*

In this path, one molecule of glucose from the chain of glucose molecules in glycogen is directly phosphorylated to glucose-1-phosphate by the enzyme 'phosphorylase a'. Glucose-1-phosphate is isomerised to glucose-6-phosphate, which goes for anaerobic glycolysis resulting in the production of lactic acid.

Out of the two paths, the amylolytic path is the principal path in fish whereas in mammals the glycolytic path predominates.

(b) Glycolysis

In this process glucose is hydrolysed to produce energy. It can occur in two paths as given below.

(*i*) *Aerobic Glycolysis*

In aerobic glycolysis, which is usually called 'glycolysis', each glucose molecule is degraded to two pyruvic acids with a net production of 8 ATP (Figure 3). The two pyruvic acids are further converted to two acetyl CoA with production of 6 ATP. These acetyl CoA enter into the Krebs cycle where they are oxidised in presence of oxygen to CO_2 and H_2O with production of 24 ATP (12 ATP from each acetyl CoA). Thus, each glucose, through glycolysis and Krebs cycle produces 38ATP (8 from glycolysis + 6 from conversion of pyruvic acid to acetyl CoA + 24 from Krebs cycle).

(*ii*) *Anaerobic Glycolysis*

Here each glucose is first degraded to two pyruvic acids with a net production of 8 ATP. The two pyruvic acids, in absence of oxygen (anaerobic condition), are converted to two lactic acids with consumption of 6 ATP. Thus, each glucose, through anaerobic glycolysis produces 2 ATP (8 ATP produced–6 used in the conversion of pyruvic acid to lactic acid).

Out of the two paths, glucose is broken down through aerobic glycolysis in the tissues of living fish in resting condition as they are supplied with oxygen. But after death, glucose is broken down through anaerobic glycolysis as the tissues are no longer supplied with oxygen and anaerobic condition prevails in the tissues. Anaerobic glycolysis also takes place in the tissues of living animals including fish while executing high muscular activity, for example, in the skeletal muscle while fish is swimming fast.

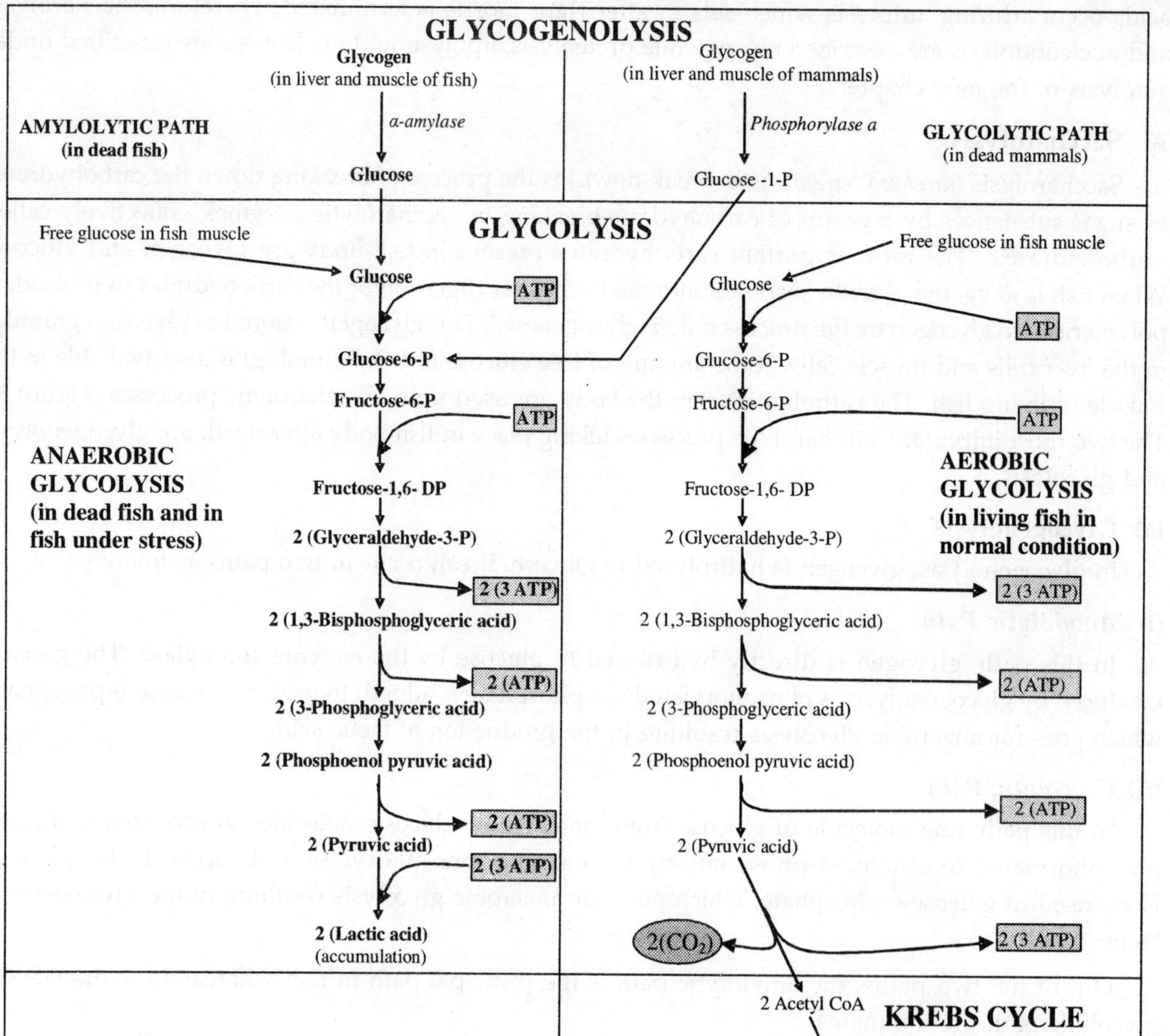

Figure 3: Break down of tissue carbohydrates in dead fish (left) and in living fish (right)

Thus, after death of fish, glycogen is hydrolysed to glucose through the amylolytic path and subsequently this glucose undergoes anaerobic glycolysis with production of two lactic acids and two ATP from each glucose. The lactic acid thus produced accumulates in the fish tissue and lowers its pH. This low pH retards the growth of most of the microbes, thereby hindering the microbial spoilage of the fish. This is the reason why rigor mortis, which runs concomitant with acidification of muscle, retards spoilage of fish for a certain length of time and is beneficial for retaining the freshness of fish. All the enzymes involved in the break down of glycogen to lactic acid such as phosphorylase a, α-amylase and the enzymes of the aerobic and anaerobic glycolysis path are included under carbohydrases.

B. Nucleotidolysis

Nucleotidolysis (*nucleotido*: of nucleotide; *lysis*: break down) or nucleotide degradation is the process of break down of nucleotides by a group of enzymes, collectively called 'nucleotidases'. The nucleotides are the building blocks of nucleic acids (RNA and DNA). The nucleotides, which build up the nucleic acids, are adenosine monophosphate (AMP), guanosine monophosphate (GMP), cytidine monophosphate (CMP), thymidine monophosphate (TMP) and uridine monophosphate (UMP). Besides these there are also other nucleotides present in the tissue. Among them, the most important with respect to rigor mortis is adenosine triphosphate (ATP). It is a nucleotide as it contains all the three components possessed by a nucleotide such as i) a heterocyclic purine or pyrimidine base, ii) a 5-carbon sugar and iii) one or more phosphate groups (Figure 4). ATP has the heterocyclic purine, adenine, the 5-carbon sugar, D-ribose and three phosphate groups. When fish is alive, ATP is used as the 'energy currency' of the cells of fish tissue. It is an energy-rich compound, which hydrolyses to liberate energy to be utilised for various metabolic activities of different tissues.

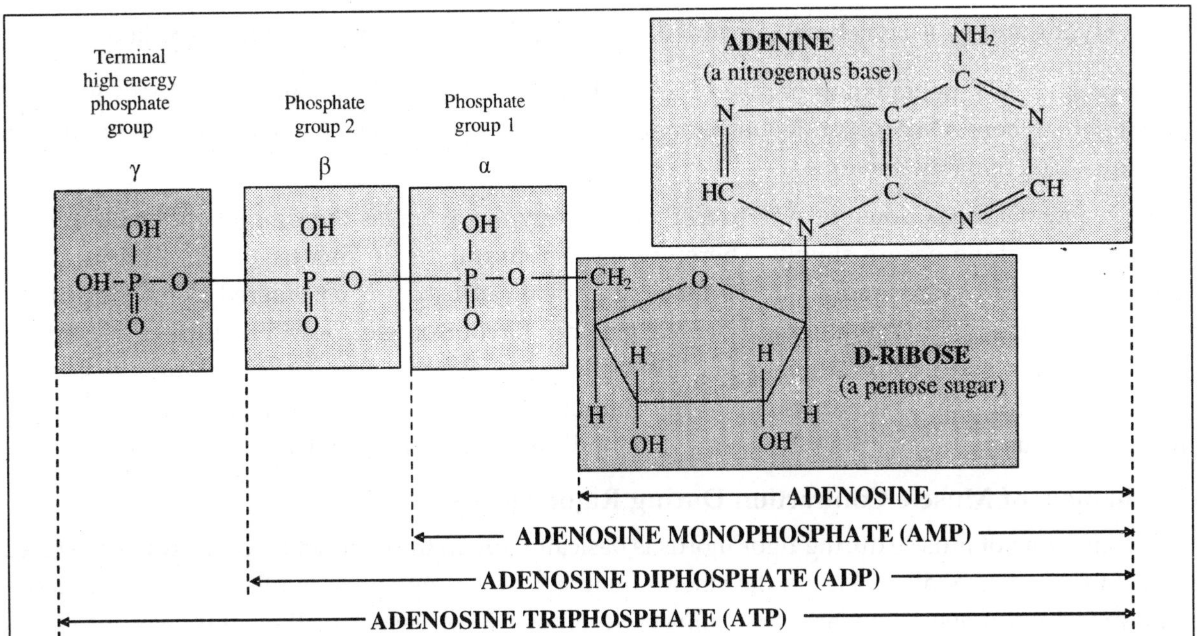

Figure 4: Structure of ATP (a nucleotide)

Degradation of ATP continues irreversibly after death of fish, due to the following two reasons.

1. Most of the cells of fish tissue remain active for sometime after death in which metabolic processes continue. Therefore, these tissue need energy, which is obtained from hydrolysis of ATP.
2. The activity of the ATP-breaking enzyme, ATPase, is enhanced due to the fall in pH of fish tissue after death resulting from the accumulation of lactic acid. This is because low pH favours ATPase activity.

Irreversible break down of ATP in tissue of dead fish occurs in a sequence of steps as given below (Figure 5).

1. ATP, in presence of water, undergoes hydrolytic dephosphorylation (removal of phosphate group in presence of water) to produce adenosine diphosphate (ADP) and phosphoric acid ($H_3PO_4 = P_i$ or inorganic phosphate) with liberation of large amount of energy. The reaction is catalysed by the enzymes ATPase, which is a ribonuclease (phosphodiesterase/5' nucleotidase). It acts only in presence of Ca^{++} ions.
2. ADP undergoes further hydrolytic dephosphorylation to produce adenosine monophosphate (AMP or adenylic acid) and P_i. The reaction is also catalysed by a ribonuclease (phosphodiesterase/5' nucleotidase) in presence of Ca^{++} ions.
3. AMP, in presence of water, undergoes hydrolytic deamination (removal of amino group, $-NH_2$, in presence of water) to produce inosine monophosphate (IMP or inosinic acid) and ammonia (NH_3). This reaction is catalysed by the enzyme deaminase.
4. IMP, in presence of water, undergoes hydrolytic dephosphorylation to produce inosine (HxR) and P_i.

5a. Inosine, in presence of water, is hydrolysed by the enzyme ribohydrolase (nucleosidase) to hypoxanthine (Hx) and D-ribose.

5b. Inosine can also be degraded in another path to produce hypoxanthine and ribose-1-phosphate. However, the former path resulting in the formation of hypoxanthine and D-ribose is the principal one since free D-ribose is found more abundantly in tissue of dead fish than ribose-1-phosphate.

6. Hypoxanthine is sometimes converted to uric acid by the enzyme xanthine oxidase.

The path of ATP degradation in case of invertebrates such as molluscs and crustaceans is slightly different in that, here AMP is first dephosphorylated to adenosine, which is then deaminated to inosine. Rest of the steps remains the same.

The sequential break down of ATP to IMP occurs very fast whereas the subsequent break down of IMP to hypoxanthine occurs slowly. Therefore, IMP accumulates in fish tissue. It is the first ATP degradation product to accumulate in the tissue after death of fish. It is considered to have distinctive flavour and thus contributes to the characteristic pleasant flavour of fish. However, further degradation of IMP to inosine and hypoxanthine is not desirable as the pleasant flavour is gradually lost due to degradation of IMP. The concentration of hypoxanthine, inosine and other ATP degradation products in fish tissue is used to determine 'K value' which indicates the freshness of wet fish (Text box 1).

2. Mechanism of Muscle Contraction During Rigor Mortis

The stiffening of muscle during rigor mortis is basically due to the contraction of the traverse skeletal muscle. Stiffening also takes place in the sinews (tendons *i.e.* strong cords, made of connective tissues, joining muscles to bones) thus affecting connective tissues. The reason for the contraction of skeletal muscle after death was not known for a long time. Some of the earlier theories proposed to explain the mechanism of muscle contraction during rigor mortis as well as the modern theory explaining the mechanism are described below.

A. Riesser Theory (1925)

B. Fluckiger Theory (1937)

C. Embeden Theory (1925)

D. Szent-Gyorgyi Theory (1945)

E. Modern Theory

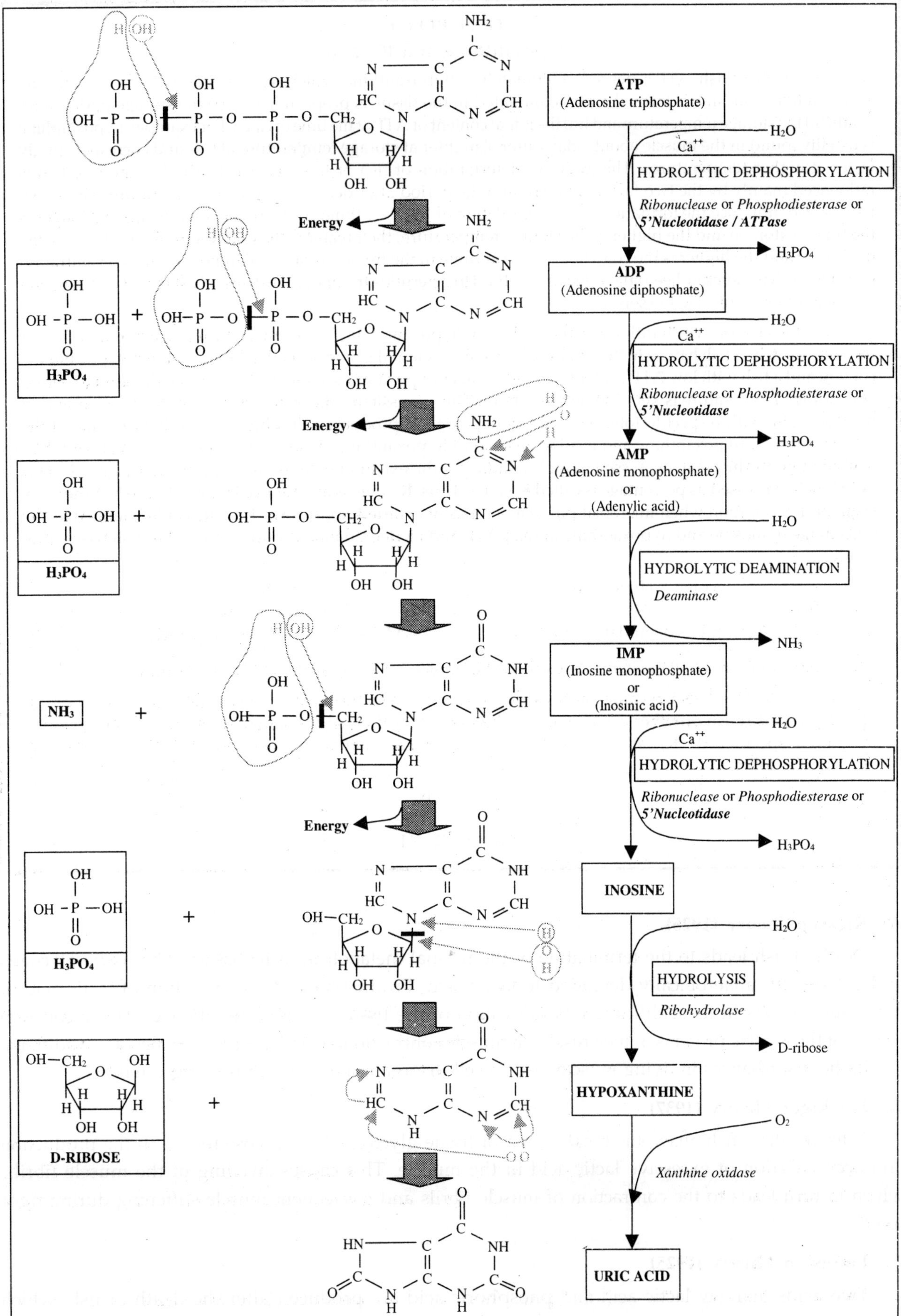

Figure 5: Post mortem degradation of ATP

TEXT BOX 1
Hypoxanthine and K value

After death of fish, ATP is degraded ultimately to hypoxanthine, which accumulates in the tissue. The rate at which hypoxanthine is formed and accumulated in the tissue is proportional to (i) the storage period after death, (ii) the storage temperature and (iii) the initial content of ATP at the time of catch. For example, hypoxanthine is usually found in the muscle about 2 days after slaughter at storage temperature of 0°C and correspondingly faster at higher temperatures. The extent of deterioration of fish or in other wards, the freshness of fish is influenced mainly by the two factors namely, storage period after death and storage temperature. These two factors also influence the formation and accumulation of hypoxanthine in fish tissue after death. The longer is the time of storage and the higher is the storage temperature, the greater is the deteriorations (*i.e.* lesser is the freshness) and the higher is the content of hypoxanthine in the tissue. Analysis of hypoxanthine concentration in fish tissue has therefore been suggested as a method to determine freshness of fish in early stages of deterioration before microbial putrefaction sets in.

The third factor, which influences the content of hypoxanthine after death, is the initial content of ATP at the time of catch. A fish having higher ATP content in its tissue at the time of death will show higher hypoxanthine content and a fish with low ATP content will show lower hypoxanthine content even though the storage period and storage temperature is kept the same for both. This necessitates the estimation of the relative amount of hypoxanthine with respect to total ATP and its degradation products. It has been found that the ratio of the concentration of hypoxanthine and inosine (instead of hypoxanthine alone) to that of the total ATP and ATP degradation products (all expressed in moles) indicates the extent of deterioration and freshness of fish. This ratio when expressed as percentage is called K value. Thus, K value is an index of freshness of fish in the early stage of deterioration when microbial putrefaction has not started noticeably. 'K value can be defined as the percentage of inosine and hypoxanthine in total ATP and its degradation products present in fish tissue after death'.

$$\text{K value} = \frac{\text{Inosine + Hypoxanthine}}{\text{ATP and its degradation products}} \times 100 = \frac{\text{HxR + Hx}}{\text{ATP + ADP + AMP + IMP + HxR + Hx}} \times 100$$

where, concentrations of all the items are expressed in moles, HxR= Inosine and Hx= Hypoxanthine.

In extremely fresh fish, K value can be as low as zero, in moderate quality fish it can be 10 to 20 and it can go upto 90 in spoiled fish. Another value suggested because of the diversity of patterns of nucleotide degradation in different species is the K_1 value, which does not take into account the content of ATP, ADP and AMP. It is expressed as follows:

$$K_1 \text{ value} = \frac{\text{HxR + Hx}}{\text{IMP + HxR + Hx}} \times 100$$

A. Riesser Theory (1925)

Death of fish leads to the termination of the normal metabolism of its tissues. The tissue glycogen and glucose are anaerobically degraded to lactic acid. Production and accumulation of lactic acid in the tissues results in its acidification. Acidification of the tissues leads to the damage of the colloidal nature of the muscle proteins, which results in muscle contraction during rigor mortis. Besides, additional metabolic disturbances in dying muscle lead to cell disorganisation and thus to rigor mortis.

B. Fluckiger Theory (1937)

After death of fish anaerobic break down of tissue glycogen and glucose results in the production and accumulation of excessive lactic acid in the muscle. This causes swelling of the muscle fibrils, which in turn leads to the contraction of muscle fibrils and consequent muscle stiffening during rigor mortis.

C. Embeden Theory (1925)

Two acids, namely lactic acid and phosphoric acid are produced after the death of fish, which accumulate in its tissues. Lactic acid is produced by anaerobic break down of glucose and phosphoric acid is produced from lactacidogens, which was supposed to be a transition product from glycogen break down. While phosphoric acid induces a rapid contraction of the muscle during rigor mortis, the lactic acid chiefly controls the duration of the contracted condition.

D. Szent-Gyorgyi Theory (1945)

The most detailed study of the mechanism of muscle contraction during rigor mortis was carried out by Szent-Gyorgyi. His findings are largely still valid although certain additions have been made to describe it in details. After death of fish, complex biochemical changes take place in the muscle, which result in the contraction of the muscle and stiffening during rigor mortis. The mechanism of muscle contraction during rigor mortis is similar to that of contraction of motor muscles during life. When a fish dies, the organic compounds in its muscle are hydrolysed by the tissue enzymes, which remain active even after death. In the initial stage, glycogen is anaerobically hydrolysed to lactic acid. Accumulation of lactic acid in the muscle reduces its pH. This in turn, stimulates the enzymes that hydrolyse the organic phosphates. The first phosphate to be hydrolysed is creatine phosphate (CrP or phosphocreatine) producing creatine (Cr) and phosphoric acid (P_i). The next phosphate to be hydrolysed is ATP producing ADP and Pi. Hydrolysis of ATP leads to its gradual disappearance from the muscle and thereby its concentration in the muscle falls.

In living fish, the main muscle proteins, actin and myosin, remain in their dissociated form by ATP. Disappearance of ATP in the muscle after death, leads to the binding of actin and myosin forming 'actomyosin complex'. This is accompanied by a change in the colloidal state of the proteins leading to myofibrillar contraction and a corresponding straining of the muscle, thus producing rigor mortis. Microscopic investigation of fish muscle shows that in a pre-rigor fish in which the muscle has not yet stiffened, the muscle fibers (*i.e.* muscle cells or myofibers) are tightly pressed together and the elements of connective tissue are closely wedged between them. But once stiffening starts after death, the myofibrillar proteins (actin and myosin) swell up forming Cohnheim's fields and the lateral striations (A and I bands) disappear. Once rigor mortis is terminated, the muscle becomes soft again and the lateral striations reappear, but the diameter of the muscle fibers become smaller and the gaps become filled with lyophilic fluid diffusing from the muscle cells through the sarcolemma. During rigor mortis the muscle remains stiffened. After sometime, rigor mortis is terminated and the muscle regains the original softness. The reason of the stiffened muscle becoming soft again, though not clearly understood, has been described to have been connected with the deformation of protein molecules and their reduced capacity to form actomyosin complex.

E. Modern Theory

The modern theory explains the mechanism of muscle contraction at molecular level. According to the modern theory, it is the disappearance of ATP from the muscle after death, which is responsible for the muscle contraction; not the fall in pH resulting from the production and accumulation of lactic acid formed by the anaerobic break down of tissue glycogen and glucose after death. The modern theory is also called the 'sliding filament-cross bridge theory'. According to this theory, due to the disappearance of ATP, the actin filaments form cross bridges (bonds) with myosin filaments after which the actin filaments slide past the myosin filaments towards the center of the sarcomere resulting in the contraction of the muscle fibers, which leads to stiffening of the muscle during rigor mortis.

Thus the modern theory is mainly based on the theory given by Szent-Gyorgyi (1945). However, the exact mechanism of how ATP retains actin and myosin in their dissociated form and exactly how muscle contracts due to formation of actomyosin was not described by Szent-Gyorgyi. The modern theory clearly delineates the exact mechanism of muscle contraction during rigor mortis, which is described below in details.

When a fish dies, its neuroendocrine system, which controls all the activities going on inside the body of live fish, ceases to function. Due to this, respiration stops leading to decrease in the oxygen content of blood. Simultaneously heart beat stops and blood is no more circulated through the tissues.

As oxygen content of blood decreases and blood circulation stops, the tissues no longer get oxygen. Thus anaerobic condition prevails in the fish tissues after death. But most of the tissues remain alive for sometime after death where metabolic processes continue. Therefore, these tissues need energy, which is obtained from the hydrolysis of ATP to ADP and P_i. Besides, hydrolysis of ATP is also enhanced, as the activity of the ATP-hydrolysing enzyme, ATPase, is favoured due to the fall in pH of the tissues resulting from the accumulation of lactic acid. The hydrolysis of ATP would lead to a fall in the ATP concentration in the tissues. But the tissues try to maintain the concentration of ATP at a constant level. For a certain lapse of time, fish tissues can maintain the constant ATP level. This is accomplished in two ways as given below.

(*a*) Hydrolysis of phosphocreatine (Figure 6) and

(*b*) Anaerobic oxidation of glucose (Figure 3)

(a) *Hydrolysis of Phosphocreatine (PC)*

Before ATP concentration falls below the original level, phosphocreatine (PC), which is present in the cells at a concentration 3-4 times greater than that of ATP, is hydrolysed by the enzyme creatine kinase to creatine (Cr) and P_i with liberation of energy (Figure 6). Utilising this energy, ATP is synthesised from ADP obtained from hydrolysis of ATP and Pi obtained from hydrolysis of PC. Thus ATP level is maintained at the cost of PC. Therefore, the concentration of PC goes down in the tissue after death. The energy liberated from hydrolysis of one molecules of PC can synthesise one molecule of ATP. Hydrolysis of PC continues till its concentration reaches 60 per cent of the original concentration, below which it does not hydrolyse. Unlike fish in which ATP is synthesised at the cost of PC to maintain a constant level of ATP in the tissue, in other types of meat, an enzyme named myokinase suppresses the break down of ATP thereby maintaining a constant level of ATP.

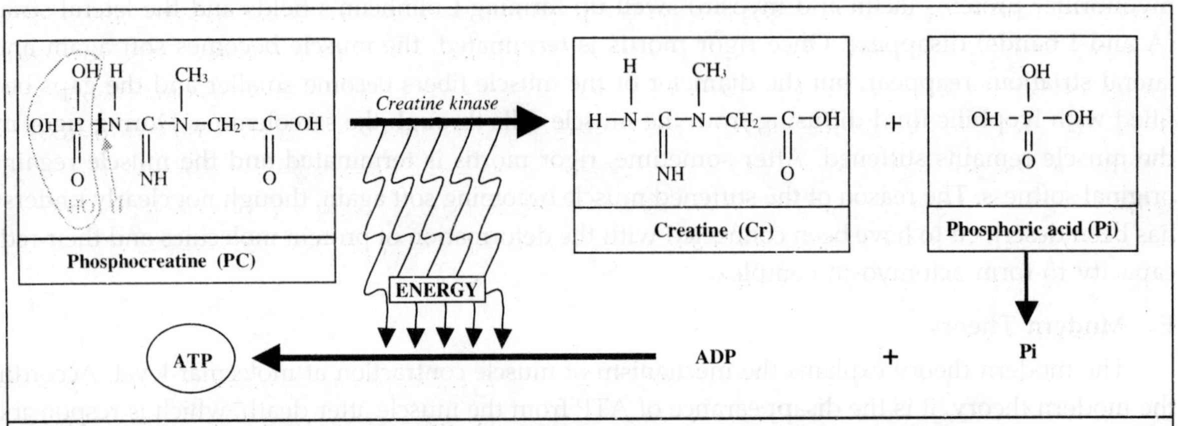

Figure 6: Hydrolysis of phosphocreatine and synthesis of ATP

(b) *Anaerobic Oxidation of Glucose*

The muscle cells of fish contain reserve glycogen granules. Some amount of free glucose (0.1-3.9μ mol/g) is also available in the muscle of live fish. After death of fish, the reserve glycogen is hydrolysed to glucose by the enzyme α-amylase through the amylolytic path. The free glucose present in the tissue as well as that produced by amylolytic hydrolysis of glycogen is anaerobically oxidised to lactic acid as supply of oxygen to the tissue stops and anaerobic condition prevails in the tissue after death. Anaerobic oxidation of each molecule of glucose produces 2 lactic acid and 2 ATP molecules (Figure 3). Unlike in live tissue, this reaction is irreversible in dead fish (Text box 2). The ATP produced replenishes the ATP utilised for the residual metabolic processes continuing after death. Production and accumulation of lactic acid in the tissue leads to a fall in tissue pH from near neutral to 6.2-6.5. Fall in pH stimulates the enzyme ATPase that hydrolyses ATP to ADP and Pi. In case of fish there is a limiting pH below which break down of glucose to lactic acid stops, but in case of other animals, break down of glucose continues till all the glucose and glycogen are completely exhausted.

TEXT BOX 2
Pattern of ATP utilisation in live fish

ATP is the energy currency of cells. The cells of fish body get energy for their metabolic processes from hydrolysis of ATP to ADP and Pi. The pattern of ATP utilisation in dead fish is different from that in live fish in normal condition and live fish under stress. In dead fish, initially ATP hydrolysis is reversible, but afterwards it becomes irreversible, but in live fish it is always reversible. For comparison among the three, the patterns of ATP utilisation in live fish are described below.

(1) Live fish in normal condition,

(2) Live fish under stress

(1) Live Fish in Normal Condition

In live fish, the cells try to maintain a constant level of ATP concentration. This necessitates the replenishment of the hydrolysed ATP through synthesis of fresh ATP. In normal condition, when fish is not executing high muscular activity, supply of oxygen to the cells is adequate. In presence of oxygen, glucose is aerobically oxidised through the long path of glycolysis and Krebs cycle to CO_2 and H_2O with liberation of large amount energy. Trapping this energy, ATP is synthesised from ADP and P_i produced from hydrolysis of ATP. The energy liberated from aerobic oxidation of each glucose molecule synthesises 38 ATP molecules. Thus 38 ATP molecules are produced from aerobic oxidation of one glucose molecule. Therefore, if 38 ATP are hydrolysed to get energy for a particular metabolic process, aerobic oxidation of only one glucose molecule can replenish all the 38 ATP consumed in the process, thus bringing the concentration of ATP back to the original level. In this way, the concentration of ATP in live fish in normal condition is maintained by the aerobic oxidation of glucose. Thus, here ATP hydrolysis is always reversible.

(2) Live Fish Under Stress

A live fish is said to be under stress when it executes high muscular activity to chase a prey or to escape from a predator or from a gear. Due to the high muscular activity, the metabolic processes of the muscle cells increase due to which they need large number of ATP. Thus the demand for ATP increases suddenly due to the short bursts of extreme muscular activity. Oxygen, by the usual supply through respiration and circulation, cannot reach the muscle cells so quickly as to oxidise glucose aerobically. Moreover, aerobic oxidation takes place through the long and complex path of glycolysis and Krebs cycle and therefore is not a process quick enough to produce ATP so as to meet the suddenly increased demand for ATP. Therefore, the concentration of ATP cannot be maintained at a constant level by aerobic oxidation of glucose. In this situation, the ATP level is maintained in two ways as given below.

(*a*) Hydrolysis of phosphocreatine and

(*b*) Anaerobic oxidation of glucose

(*a*) Hydrolysis of Phosphocreatine

Phosphocreatine (PC) or creatine phosphate is an energy rich phosphate compound, which has similar function as ATP. It is a reserve form of high-energy phosphate compound. Its concentration in the muscle cells is about 3-4 times greater than that of ATP. Its function is to maintain ATP concentration in the muscle cells at a constant high level, particularly in skeletal muscle cells, which must perform intermittent and somewhat strenuous work. It hydrolyses to creatine (Cr) and P_i liberating energy. When ATP hydrolyses to ADP and P_i to provide energy for the metabolic activity of the highly active muscle cells, its concentration in the cells would go down. But the concentration of ATP is maintained at a constant level by replenishment of the ATP utilised, through synthesis of fresh ATP. Before ATP concentration falls below the original level, PC in the cells is hydrolysed by the enzyme creatine kinase to Cr and P_i with liberation of energy. Utilising this energy, ATP is synthesised from ADP obtained from hydrolysis of ATP and Pi obtained from hydrolysis of PC. Thus ATP level is maintained at constant level at the cost of PC. Therefore, the concentration of PC goes down in the tissue during stress. The energy liberated from hydrolysis of one molecule of PC can synthesis one molecule of ATP.

(b) Anaerobic Oxidation of Glucose

In anaerobic oxidation of glucose, the glucose produced by glycogenolysis of the reserve glycogen granules present in the liver and muscle cells as well as the free glucose present in the muscle cells is broken down in absence of oxygen through a short quick process of anaerobic glycolysis to lactic acid with liberation of energy. Each molecule of glucose is broken down to 2 molecules of lactic acid. The energy produced from each glucose molecule can synthesise only 2 ATP molecules as compared to the 38 ATP molecules produced by aerobic oxidation of glucose. Thus anaerobic oxidation, though inefficient producing only 2 ATP molecules per glucose

Contd...

Text Box 2–Contd...

molecule, occurs during stress, because of two reasons, such as (1) Oxygen supply during stress is insufficient for aerobic oxidation of glucose and (2) Anaerobic oxidation is a process short and quick enough to meet the suddenly increased ATP demand. Due to the anaerobic oxidation of glucose to lactic acid, the concentration of glucose falls and that of lactic acid goes up in the tissue resulting in the fall in tissue pH.

When the stress disappears, the fish rests and recovers. During this recovery period, it breathes intensely and oxygen is consumed in large quantity. The amount of oxygen consumed gradually diminishes until respiratory rate returns back to the normal level. With the help of this large supply of oxygen, glucose is now aerobically oxidised through the long complex path of glycolysis and Krebs cycle to CO_2 and H_2O with liberation of large amount of energy. The energy liberated from one molecule of glucose synthesises 38 ATP molecules. Thus, large number of ATP is produced by aerobic oxidation of glucose during recovery. These ATP are hydrolysed to ADP and Pi with liberation of energy. This energy is utilised for reversing the processes that takes place during stress, so that the tissue comes back to the normal condition. The energy from ATP hydrolysis is utilised for

(*a*) Synthesis of phosphocreatine and

(*b*) Synthesis of glucose.

(*a*) Synthesis of Phosphocreatine

The creatine produced from hydrolysis of PC during stress combines with the P_i produced from ATP hydrolysis trapping the energy from ATP hydrolysis to synthesise PC. As more and more PC is synthesised, its concentration comes back to the original level, as it should be in the normal condition. One molecule of ATP has to be hydrolysed for the synthesis of one molecules of PC.

(*b*) Synthesis of Glucose

The energy produced from the hydrolysis of ATP during recovery is used to convert back lactic acid to glucose inside the liver. Due to this, the concentration of lactic acid in the tissue falls, that of glucose goes up to the original level and the pH of tissue comes back to the neutral range. As anaerobic oxidation of each molecule of glucose produces 2 molecules of lactic acid and 2 molecules of ATP during stress, in the reverse process taking place during recovery, 2 lactic acid utilise 2 ATP to synthesise each molecule of glucose. Therefore, the 38 ATP produced from aerobic oxidation of one glucose molecule can convert back 38 molecules of lactic acids to 19 molecules of glucose.

In this way, the level of ATP in the tissue is maintained for a certain lapse of time after death. But after sometime, when the concentration of PC goes down below 60 per cent of its original concentration, it no longer hydrolyses to provide energy for the synthesis of fresh ATP. Similarly, anaerobic oxidation of glucose to lactic acid stops upon reaching the limiting low pH and it no longer provides energy for the synthesis of fresh ATP. Therefore, after this period during which ATP break down is reversible, irreversible break down of ATP starts. In this process small amount of energy is liberated which is utilised for the residual metabolism. The irreversible break down of ATP, which has already been described under nucleotidolysis, leads ultimately to the formation of hypoxanthine and D-ribose.

Disappearance of ATP and Muscle Contraction

Most of the membranes of the muscle cells of fish such as sarcolemma[3] and membrane of the sarcoplasmic reticulum[4], maintain a concentration gradient of different ions across them (Figure 7). If a particular ion, say Ca^{++} is present at a low concentration on one side of the membrane, a high concentration of the ion is maintained on the other side. As these membranes are semipermeable, ions can freely diffuse through them downhill along the concentration gradient from the high concentration side to the low concentration side till equilibrium is attained and the concentration of the ion on both the sides becomes equal. To maintain the concentration gradient, these ions have to be pumped back uphill against the concentration gradient from the low concentration side to the high concentration side. This uphill transfer of ions against concentration gradient is done by specific ion pumps. Each pump is specific for a particular ion; say Ca^{++} pump to pump only Ca^{++} ions, Na^+ pump to pump only Na^+ ions. These pumps require energy to transfer ions uphill. This energy is obtained from ATP. It is a

3 Sarcolemma: Plasmalemma or cell membrane of muscle cells.

4 Sarcoplasmic reticulum: Endoplasmic reticulum of muscle cells.

type of active transport. Thus in live fish, the concentration gradient of ions is maintained across the membranes by utilising energy obtained from hydrolysis of ATP.

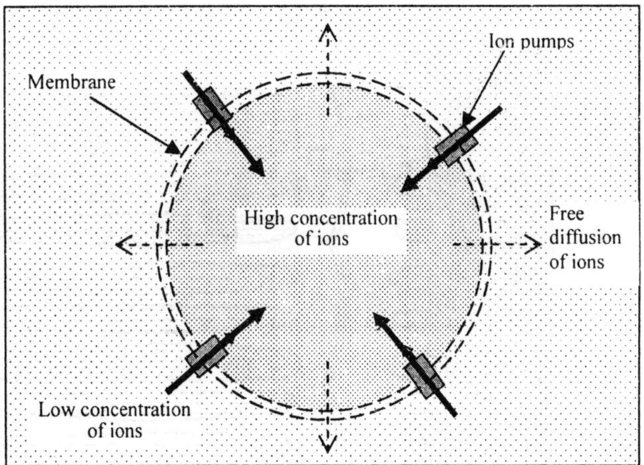

Figure 7: Ion pumps on the membranes

In the muscle cells of live fish (Figure 8), the two lateral tubules of each T system (triad system), which are the transverse tubules of the sarcoplasmic reticulum, maintain a higher concentration of Ca^{++} ions within them as compared to the low concentration of Ca^{++} ions in the sarcoplasm and the myofibrils. The membrane of the lateral tubules has Ca^{++} pumps to pump Ca^{++} ions from the sarcoplasm and myofibrils into the tubules so that a higher concentration of Ca^{++} ions is maintained inside the tubules. The Ca^{++} pumps get energy from ATP and pump Ca^{++} ions uphill against the concentration gradient from the sarcoplasm and myofibrils into the tubules.

After the death of fish, ATP gradually disappears due to its irreversible break down to hypoxanthine and D-ribose. When the concentration of ATP in the muscle tissue reaches a certain low level, about 5 per cent of the original concentration present in live fish, contraction of muscle tissue begins and stiffening of fish body starts. Due to the gradual disappearance of ATP, the Ca^{++} pumps present on the membrane of the lateral tubules of the T systems do not get energy and therefore cease to function (Figure 9). Now Ca^{++} ions diffuse freely through the membrane of the lateral tubules downhill along the concentration gradient from the high concentration side within the tubules to the low concentration side in the sarcoplasm and myofibrils. These Ca^{++} ions, reach the myofibrils very fast because of the spreading network of the T systems. As the T systems run along the Z lines, Ca^{++} ions are released along the Z

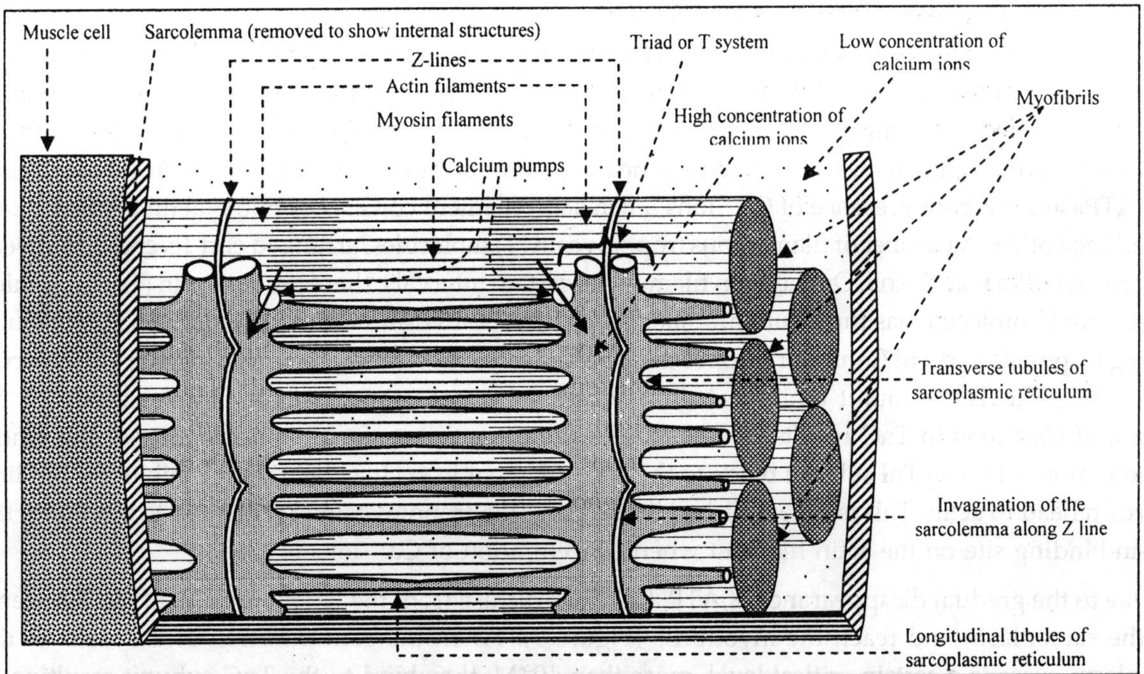

Figure 8: Muscle cell before death.
[Note the sarcoplasmic reticulum with high concentration of calcium ions (small dots). The calcium pumps located on the membrane of the sarcoplasmic reticulum constantly pump these ions into it using ATP.]

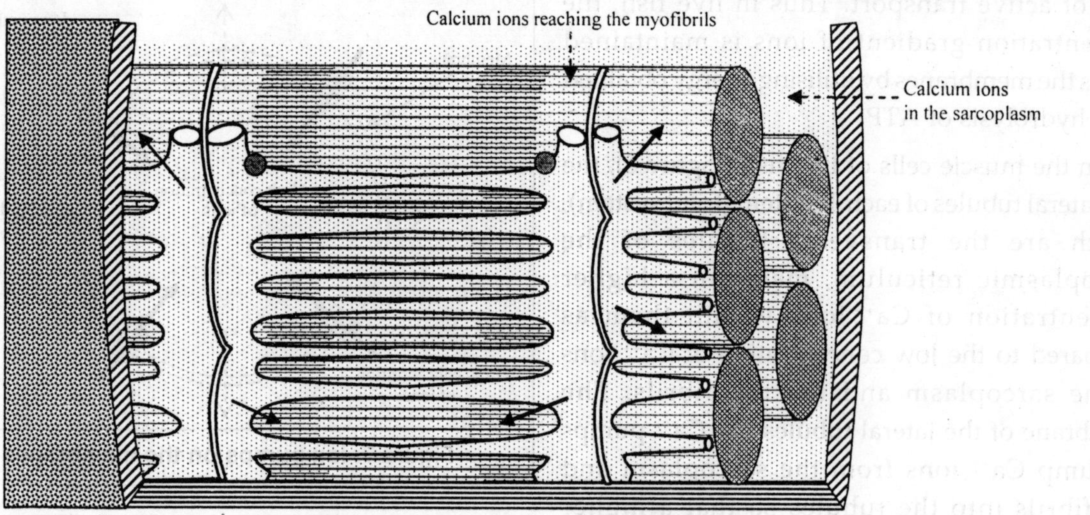

Figure 9: Muscle cell after death.
[Due to the disappearance of ATP after death, the calcium pumps cease to function. Therefore the calcium ions freely diffuse into the sarcoplasm and reach the myofibrils.]

lines. This free diffusion of Ca^{++} ions continues till equilibrium is attained and concentration of Ca^{++} ions on both the sides of the membrane becomes equal. Thus after death of fish, due to the gradual disappearance of ATP, the membrane of the lateral tubules loses its ability to maintain concentration gradient of Ca^{++} ions across it and these ions reach the myofibrils very fast.

The Ca^{++} ions reaching the myofibrils take a great role in the contraction of muscle during rigor mortis. They have a two fold effect which leads to muscle contraction. The effects are given below.

(*a*) Formation of actin-myosin cross bridge

(*b*) Production of energy for sliding of actin filament over myosin filament

(*a*) Formation of Actin-Myosin Cross Bridge

In the myofibril, the thin actin myofilaments project from the Z lines and the thick myosin myofilaments remain between the actin filaments (Figure 10). Each myosin filament consists of a bundle of myosin molecules arranged in a head-to-tail manner. Each myosin molecule has a fibrous tail, a small neck and a globular head. The myosin heads project outward towards the actin filaments and have ATPase activity in presence of Ca^{++} ions. Each actin filament consists of two F-actin strands coiled around each other. A number of filamentous tropomyosin (T) molecules, arranged end-to-end, fit between the grooves of the F-actin coil. On the actin filament, at certain intervals, there are troponin (Tn) molecules. Each troponin molecule has three subunits such as TnT (tropomyosin-binding troponin), TnC (calcium-binding troponin) and TnI (inhibitory troponin). The TnT subunit binds to the tropomyosin (T) molecule (Figure 11). The TnC subunit binds Ca^{++} ions when the concentration of Ca^{++} ions exceeds 10^{-6}M. Binding of Ca^{++} ions to TnC results in a change in the three-dimensional conformation of the whole troponin molecule. The TnI subunit binds to the actin filament in such a way that it inhibits the binding of myosin head to actin. This is accomplished by pressing the tropomyosin (T) filament to bind onto the myosin-binding site on the actin filament when concentration of Ca^{++} ions is low.

Due to the gradual disappearance of ATP, Ca^{++} ions diffuse from the lateral tubules of the T systems into the sarcoplasm and reach the myofibrils (Figure 9). When the concentration of Ca^{++} ions in the sarcoplasm exceeds a certain critical level, more than 10^{-6}M, they bind to the TnC subunit resulting in a change in the three-dimensional conformation (quaternary structure) of the globular protein troponin (Figure 11). Due to this, the linkage among TnC, TnI and TnT gets tightened and TnI gets delinked from the actin filament. The tropomyosin filament, which is bonded to TnT subunit, moves deeper into the

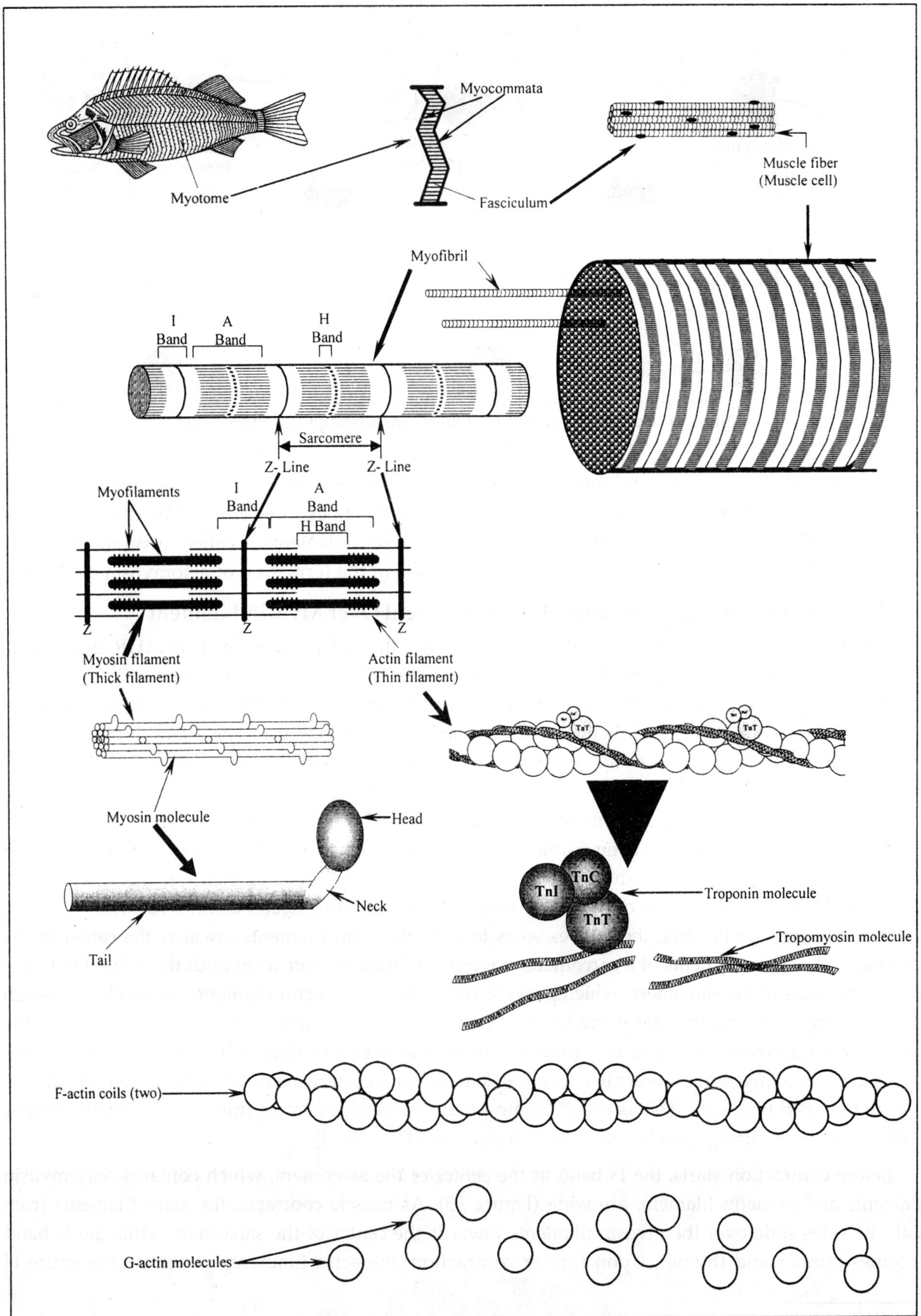

Figure 10: Structure of fish muscle

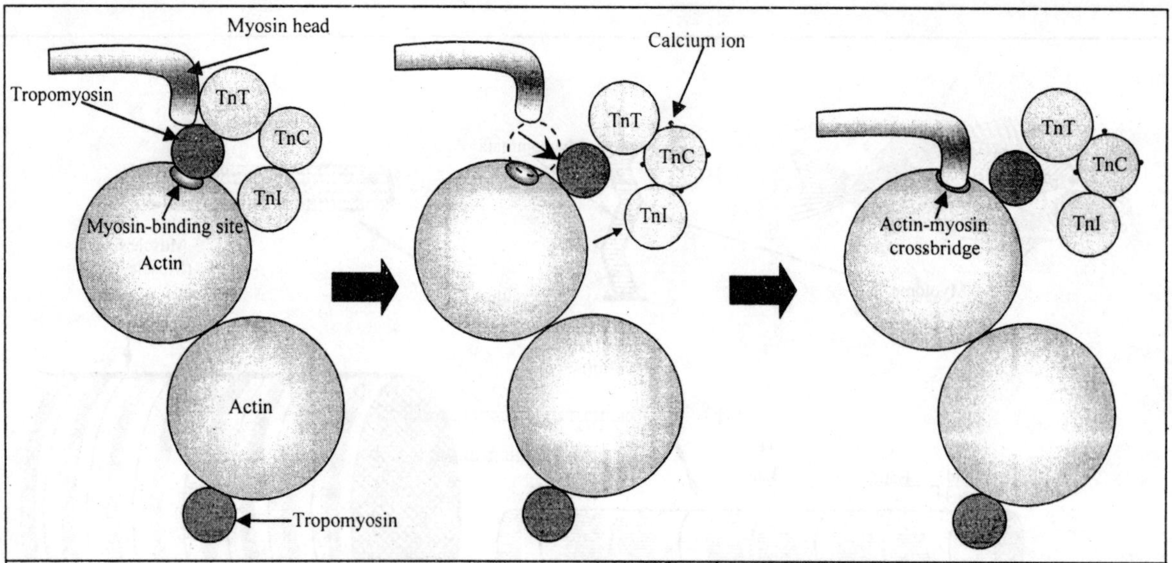

Figure 11: Formation of actin-myosin cross-bridge (Section view)

actin groove exposing the myosin-binding site on the actin filament. Now, the myosin head, projecting from the myosin filament towards the actin filament, binds to the exposed myosin-binding site on the actin filament. The bond formed between the actin and myosin filaments is called 'actin-myosin cross bridge'. The formation of actin-myosin cross bridge results in the formation of actomyosin complex.

(*b*) Production of Energy for Sliding of Actin Filament Over Myosin Filament

The myosin head has Ca^{++} induced ATPase activity. It can hydrolyse ATP to ADP and P_i with liberation of energy only in presence of Ca^{++} ions. Some ATP[5] molecules are present bound to the myosin heads. Thus both ATP and ATPase are present on the myosin heads and only Ca^{++} ions are required to hydrolyse ATP. After death of fish, Ca^{++} ions diffuse freely from the lateral tubules of the T systems and reach the myosin heads. In presence of these Ca^{++} ions, the Ca^{++} induced ATPase of the myosin heads hydrolyses ATP with liberation of energy. This energy is used to bend the series of myosin heads projecting from the myosin filaments towards the center of the sarcomere (Figure 12). As the myosin heads are bonded to the actin filaments, the actin filaments are pulled towards the center of the sarcomere. Once the energy is exhausted, the myosin heads come back to their vertical position and attach to the next binding sites on the actin filaments. The myosin heads again hydrolyse ATP and utilise the energy for bending themselves so as to slide the actin filaments towards the center of the sarcomere. As actin filaments of a sarcomere move closer to each other from both the sides, the two Z lines at the ends of the sarcomere, which provide anchorage to the actin filaments, come closer to each other, thereby reducing the sarcomere length. As the sarcomere length is reduced, the length of the muscle fibers decreases and muscle contracts. Due to contraction of muscle fish body stiffens. In this way, the series of myosin heads act like levers and make oar-like movements to slide the actin filaments over the myosin filaments. This is called 'sliding filament mechanism' of muscle contraction. Similar events also occur during muscle contraction in live fish (Text box 3).

Before contraction starts, the H band at the center of the sarcomere, which contains only myosin filaments and no actin filament, are wide (Figure 13). As muscle contracts, the actin filaments from both the sides slide over the myosin filaments towards the center of the sarcomere. Thus the H band becomes thinner and thinner. Upon further contraction, the actin filaments penetrate the entire H

---

5 These account for the 5% of the original concentration of ATP at which contraction begins and the rest 95% is hydrolysed irreversibly after death to get energy for the residual metabolic processes before contraction begins. This 5% ATP is used up in the contraction process.

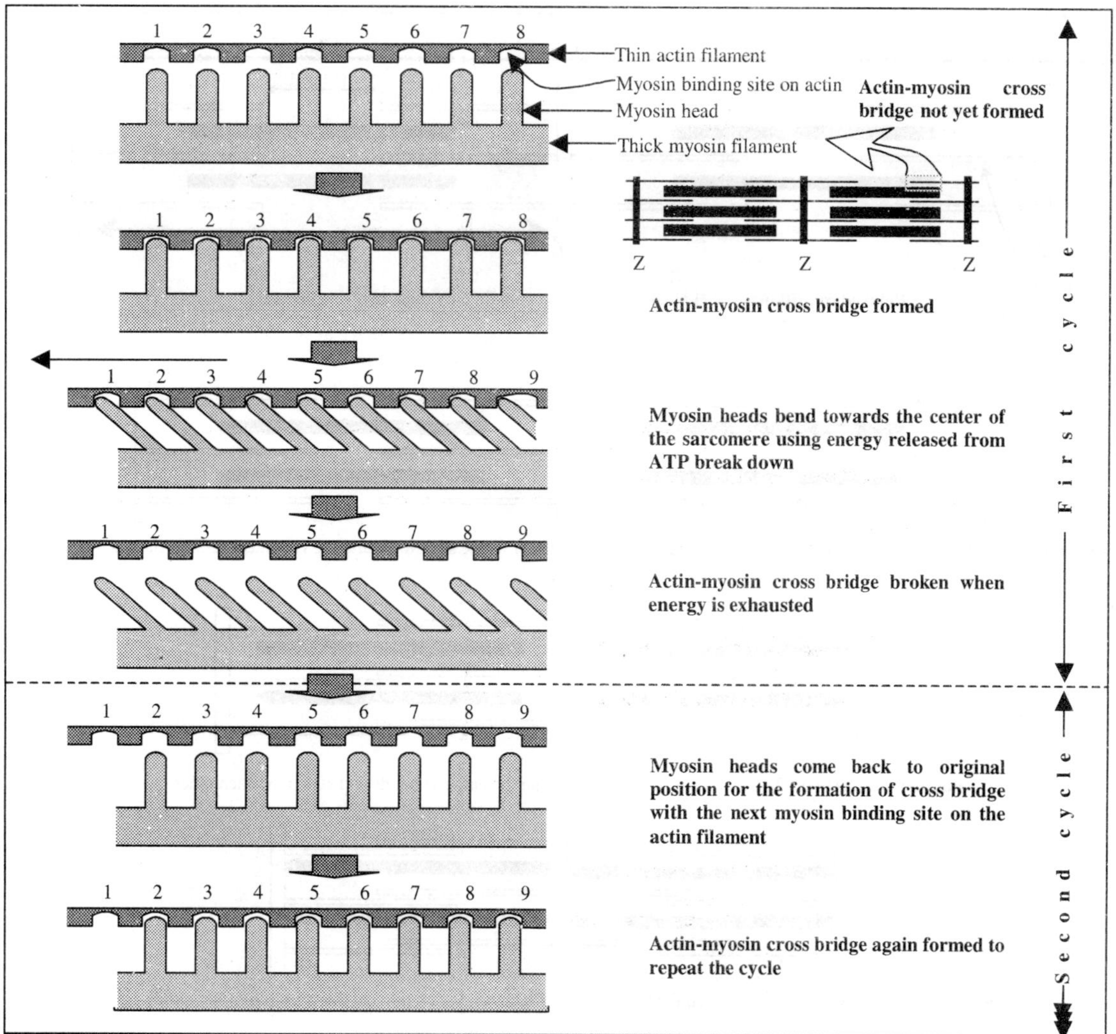

Figure 12: Movement of actin filament by myosin heads

band and slide past each other until the myosin filaments touch the Z lines on both the sides. After this no more contraction can take place. The muscle remains in the contracted state permanently as Ca^{++} ions are not pumped back to the lateral tubules of the T systems. This permanently contracted condition of muscle after death is called rigor mortis.

3. Mechanism of Softening of Muscle During Resolution of Rigor Mortis

During rigor mortis, the fish body remains stiffened. After sometime, resolution of rigor mortis starts and the fish muscle begins to soften. Rigor mortis is said to be completely terminated when the softness of fish muscle comes back to the original softness level (Figure 2). Although contraction of muscle during rigor mortis takes place due to formation of bonds between actin and myosin, in softening of muscle during resolution of rigor mortis, it is not the bonds between actin and myosin which are broken, rather the Z lines, which provide anchorage to the actin filaments are disintegrated due to their hydrolysis by the endogenous enzymes of the fish muscle. The Z lines are made up of the protein α-actinin. Towards the end of rigor mortis, autolysis sets in and the endogenous enzymes start hydrolysing the tissue components. The tissue proteins are hydrolysed by proteolytic enzymes. The most important proteolytic enzyme of fish muscle is cathepsin. As most of the proteolytic enzymes favour low pH, their activity is enhanced after death due to the low tissue pH resulting from accumulation of lactic acid. When the protein, α-actinin of the Z lines, is hydrolysed, the actin filaments lose their anchorage, which results in the softening of muscle during resolution of rigor mortis.

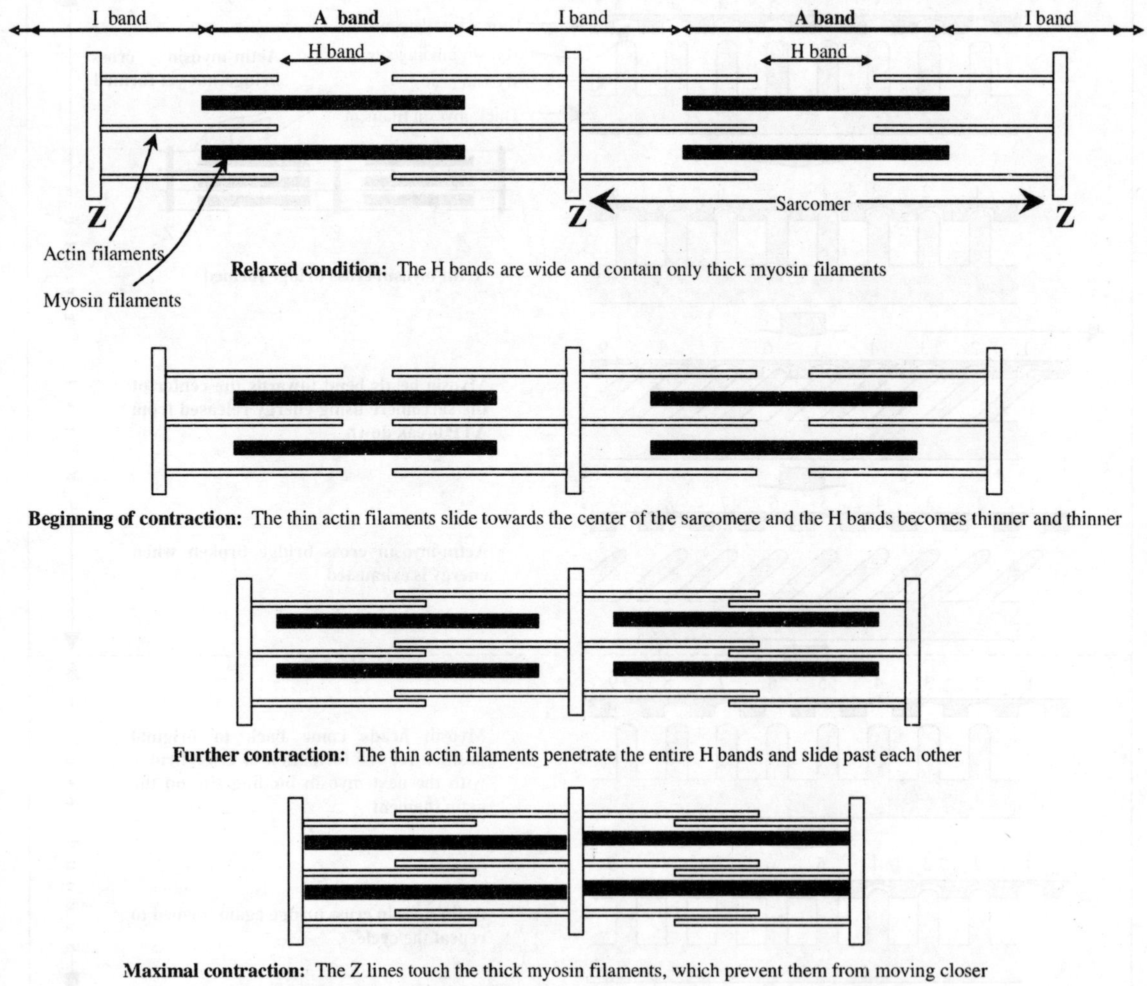

Figure 13: Sliding of filaments during muscle contraction

IV. Measurement of Rigor Mortis

Rigor mortis in fish is measured by determining the 'toughness' of the muscle. Among several methods, the direct method of assessing rigor mortis by measuring toughness of whole fish with the help of an instrument developed by Messtorff (1954) is worth mentioning (Figure 14). The instrument has a blunt-ended piston. For measurement of rigor mortis, first the instrument is kept vertically on the muscle surface just touching it. Then the piston is pushed into the flesh by applying constant pressure transferred via a compression spring until the limiting plate at the lower end touches the body surface. The depth of the imprint is measured through a small, glass-covered opening. The pressure exerted does not bring about any change in the mechanical condition of muscle of such nature as to influence the course of rigor negatively. As rigor proceeds, toughness increases and the depth of penetration of the piston decreases. The indicator scale indicates the toughness with 0 (zero) value at the top of the scale. At 0 on the scale, penetration is minimum and toughness maximum whereas at 30, penetration is maximum and toughness minimum. Toughness is measured in different locations of the fish body, say caudal area, dorsal fin area etc.

The values are expressed percentage-wise taking 100 as the 'original softness level' (the softness when fish is alive). When softness decreases with increase in toughness as rigor proceeds, this percentage goes down. Rigor is considered terminated when the softness comes back to the original softness level.

TEXT BOX 3

Contraction and relaxation of muscle in live fish

Contraction and relaxation of muscle in live fish takes place almost in a similar manner as that in dead fish except few deviations. When a fish is alive, muscle contraction and relaxation takes place by nerve impulse. The impulse is created in the brain and carried down to the muscle cells by nerve cells. The impulse is transmitted as electrical impulse by alternate depolarisation and repolarisation of the cell membrane of the nerve cells. The nerve cells connected to the muscle cells deliver the impulse at the myoneural junction (junction between muscle cell and nerve cell). At the junction, the electrical impulse is converted into a chemical impulse and a neurotransmitter called acetylcholine is secreted from the nerve endings at the myoneural junction. When the neurotransmitter reaches the muscle cells, it produces depolarisation of its cell membrane thereby exciting the cell membrane of the muscle cells. In the muscle cells, the cell membrane or sarcolemma invaginates into the cells along the Z lines forming the central tubule of the T systems and the two lateral tubules of the T systems are formed by the transverse tubules of the sarcoplasmic reticulum. Thus, the central tubule of each T system is an extension of the sarcolemma. Therefore, depolarisation and excitation of the cell membrane of the muscle cells induces the depolarisation and excitation of the central tubules of the T systems. Thus, the nerve impulse reaches the central tubules of the T systems. This impulse induces the release of Ca^{++} ions from the two lateral tubules through their Ca^{++} channels, to the myofibrils along the Z lines. Localised release of Ca^{++} ions leads to muscle contraction in a similar way as for contraction during rigor mortis. Muscle is relaxed when an impulse to relax the muscle is created in the brain and transmitted to the central tubules of the T systems of the muscle cells through the nerve cells. This impulse induces the Ca^{++} pumps present on the membrane of the lateral tubules of the T systems to utilise ATP and pump back Ca^{++} ions into the lateral tubules. Once Ca^{++} ions are pumped back, the myosin heads can no longer hydrolyse ATP for muscle contraction. Besides, Ca^{++} ions are no longer available for TnC due to which the troponin molecule regains its original conformation. This leads to the binding of the tropomyosin to the myosin-binding sites on the actin. Thus, actin-myosin bond breaks. Once the bond is broken and energy is not available for contraction, the actin filaments come back to their original position just as springs come back to their original shape when loads are removed. The Z lines move back to the original position bringing back the sarcomeres to their original length. This is reflected as muscle relaxation in living condition. Muscle contraction takes place extremely quickly; in milliseconds after the muscle cells get the impulse from the nerve cells. This needs the transmission of impulse and release of Ca^{++} ions to be extremely fast. No ordinary diffusion would be quick enough to transmit the impulse and release Ca^{++} ions throughout a muscle cell simultaneously. It is the minutely spread network of the T systems which helps in the transfer of impulse and release of Ca^{++} ions. It is more than a telegraphic connection signaling contraction and relaxation. The tubules of the T systems run along the Z lines, thus delivering the nerve impulse and resulting release of Ca^{++} ions directly at the Z lines. This leads to fast contraction of the myofibrils, which is manifested as muscle contraction. Muscle contraction and relaxation is mainly based on the concentration of Ca^{++} ions in the myofibrils. Thus whether a muscle contracts or relaxes mainly depends upon the concentration of Ca^{++} ions in the myofibrils.

The softness percentage, when plotted against time, gives a typical curve (Figure 2) called 'stiffness diagram' or 'rigor diagram'. The diagram gives idea about the following aspects.

1. Pre-rigor period or lag of initiation *i.e.*, the time lapse from the death of fish till the onset of rigor mortis. In order to find out this period, it is required to note the death point and the point of initiation of rigor. The initiation point can be noted as the point where stiffening begins. But it is confusing to find out the death point, a sharp time when the fish dies. However, Pohl (1953) defined death point as the moment after which independent movement, breathing and eye reflexes can no longer be seen (so-called partial death).
2. Time taken to reach the peak, after death
3. Total duration of rigor
4. Intensity of rigor *i.e.* whether the stiffness is more pronounced or less. If the curve is steeper, then rigor is more pronounced and if it is more flattened, than rigor is less pronounced.

V. Factors Influencing Rigor Mortis

Several factors influence rigor mortis. The three important aspects of rigor mortis, which are influenced by these factors include (1) The duration of pre-rigor, (2) The duration of rigor mortis and

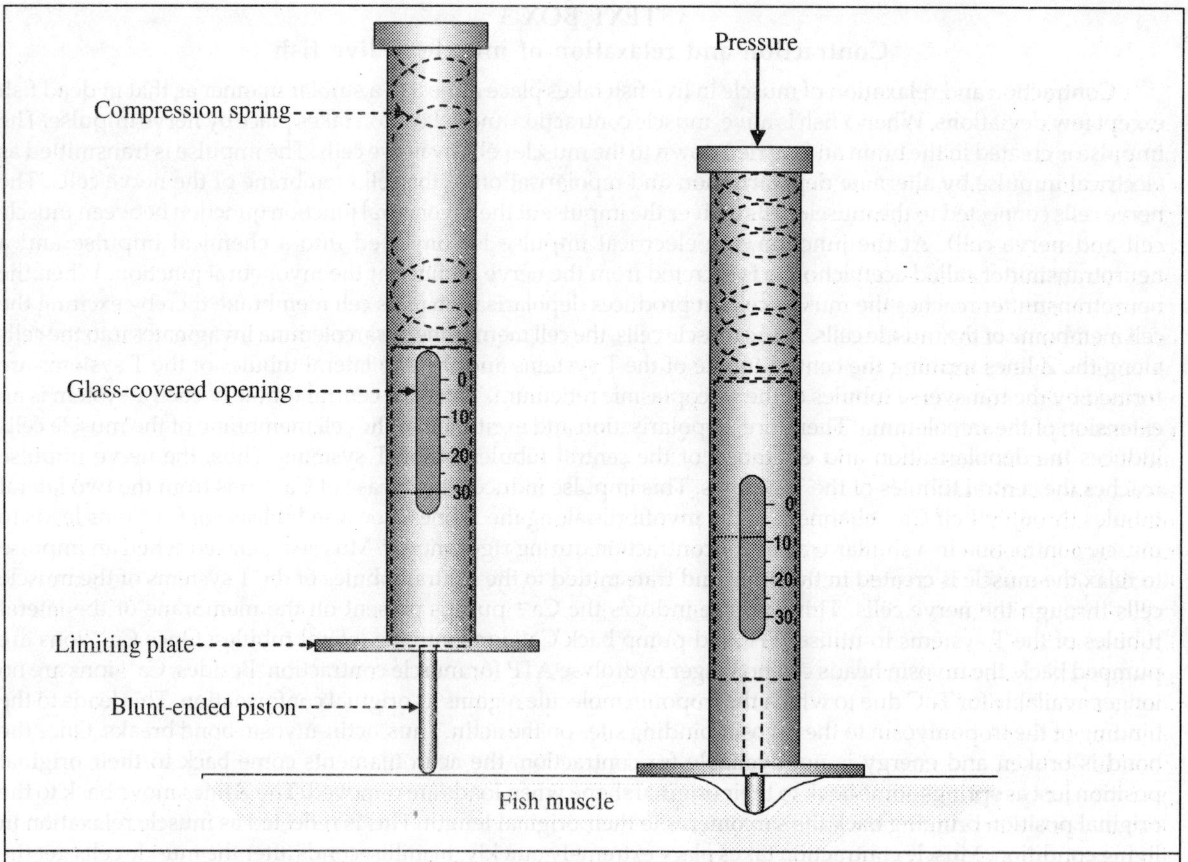

Figure 14: Measurement of rigor mortis using the instrument developed by Messtorff (1954)

(3) The intensity of rigor mortis. It is important to note that, if rigor sets in 'early', is of 'shorter duration' and 'lower intensity', it is 'poor rigor' whereas if rigor sets in 'late', is of 'longer duration' and 'higher intensity', it is 'good rigor'. Good rigor is favorable in fish processing point of view as it helps in retaining the freshness of fish for a longer duration. The factors, which influence rigor mortis are given below.

1. Intrinsic Factors

Intrinsic factors are those factors, which are inherent to the fish body and cannot be controlled by man.

(*a*) Species of Fish

(*i*) *Type of Fish*

Usually elasmobranchs (cartilaginous fishes) have poor rigor unlike teleosts (bony fishes).

(*ii*) *Body Shape*

Flat-shaped fishes show good rigor whereas round-shaped fishes show poor rigor.

(*iii*) *Activeness*

Active fishes such as herring, mackerel and pink perch show poor rigor as compared to sluggish fishes like sheat fish and carps, which show good rigor.

(*b*) Health of Fish

Diseased fishes show poor rigor as compared to healthy ones.

(*c*) Feeding Condition

In ill-fed fishes, rigor is poor while well-fed fishes show good rigor.

(d) Age and Size of Fish

In any species of fish, rigor is same for fishes of the same size and age. But smaller and low age group fish show poor rigor while larger and older fish of the same species show good rigor.

2. Extrinsic Factors

Extrinsic factors are those factors, which are not inherent to the fish body, rather are under the control of man's intentional or incidental actions.

(a) Manner of Death

Fish dies in four ways as given below.

(i) Out of asphyxia on the deck or when still held by the gear

(ii) By hitting on the head of fish

(iii) By slaughtering (beheading or decapitating)

(iv) By electroshock

Out of the four methods, fish dies in the shortest time when killed by electroshock followed by when killed by slaughtering and when killed by hitting on the head, but if fish dies out of asphyxia, death occurs after struggling for a long time. When fish is killed quickly, the onset of rigor mortis is delayed, its duration prolonged and intensity enhanced (Table 1). A long rigor mortis has protective effect on the quality of fresh fish, as it delays, for a certain length of time, the onset of the subsequent deteriorative processes of autolysis and microbial putrefaction. Therefore, when fish is killed quickly, say immediately after it is held by the gear or immediately after it is removed from water, the freshness of fish is retained for a longer period as rigor mortis sets in late, lasts longer and is of higher intensity. Thus it shows 'good rigor'. On the other hand, if fish is allowed to struggle too much for a long time and dies out of asphyxia, during struggling the glycogen present in the muscle cells is anaerobically oxidised to lactic acid very fast and ATP disappears very quickly. This leads to an early, short and less intense rigor. Due to this, the deteriorative processes of autolysis and microbial putrefaction, which follow rigor mortis, set in early after death resulting in an early spoilage and poor keeping quality of fresh fish. Thus, it shows 'poor rigor'. Therefore, fish should be killed by adopting a quick method of killing immediately after it is held by the gear and before it starts struggling too much. That is why, the most preferred method of killing fish is by electroshock followed by slaughtering and by hitting on the head. Fish should not be allowed to die out of asphyxia after struggling for long time being held by the gear.

Table 1: Influence of manner of death on duration of rigor

Fish	*Manner of Death*	*Contraction Period (hours)*	*Resolution Period (hours)*	*Total Period (hours)*
Cod	Asphyxiated	5	40	45
	Slaughtered	17	63	80

(b) Handling

If fish is handled properly without throwing around or bruising, it shows good rigor. This ultimately gives a better quality fish and longer shelf life as compared to rough-handled fish, which shows poor rigor.

(c) Storage Temperature

Fish stored at cooler temperatures shows good rigor while that stored at warmer temperatures shows poor rigor. This is one of the reasons why storage at cooler temperatures retains the freshness of fish for a longer period. Therefore, fish should always be stored at cooler temperatures immediately after harvesting.

VI. Changes During Rigor Mortis

The most important changes manifested during rigor mortis are given below.

1. Loss of texture of fish
2. Fall in pH of fish tissue

1. Loss of Texture of Fish

The texture of fish is by no means the same before and after rigor mortis. An experienced person can easily find out whether the fish is in pre-rigor, on-rigor or post-rigor stage simply by squeezing its muscle gently. Post-rigor fish shows pressure marks after such a test.

2. Fall in pH of Fish Tissue

Immediately after death, anaerobic break down of the glycogen to lactic acid starts in the muscle tissue, which continues during rigor mortis. Lactic acid is also produced before death if fish struggles too much before dying. Production and accumulation lactic acid results in a fall in pH of the muscle tissue from near neutral (7.05-7.35) to about 6.2–6.5 (Figure 48). The change in pH of fish tissue after death is as follows (Table 2).

Table 2: Change in the pH of fish tissue after death

Time after Death	*pH*
0–1 hr after catch	6.5–6.8
During rigor mortis	6.3–6.7
At the end of rigor	6.1–6.9
9 days after catch	6.8–7.0

In the warm-blooded mammalian meat (beef, pork, mutton) the pH goes down to a still lower level of 5.4 as compared to 6.2–6.5 in case of fish. This occurs due to the following reasons

1. The muscle of these animals contain a large reserve of glycogen which, through 'glycolytic hydrolysis' and subsequent 'anaerobic glycolysis' produces huge quantity of lactic acid resulting in a steep fall in tissue pH.
2. In these animals anaerobic glycolysis after death continues unabated until the whole amount of glycogen reserve in the muscle cells is exhausted.

Contrary to this, in the cold-blooded fish, pH cannot go down below certain limiting pH of about 6.2-6.5 because of the following reasons.

1. The muscle cells of fish have small reserve of glycogen.
2. There is a limiting low pH (6.2-6.5) below which anaerobic glycolysis stops as the enzymes involved in the process are inactivated below this limiting pH. No further conversion of glycogen to lactic acid takes place and pH does not go down further.

The fall in pH has both positive and negative effects. However, the gain in quality due to the positive effects is highly remarkable as compared to the loss in quality due to the negative effects. The positive effects are of greater economic importance. Fall in pH suppresses the development of spoilage microbes. Thus, the lower is the pH, the later spoilage sets in. As fall in pH is smaller in fish than that in the terrestrial animals, fish spoils much faster than the meat of these animals. That is why, fish is a highly perishable food item. The negative effect of pH fall is that at lower pH, the solubility and the degree of hydration of muscle proteins decreases, which is manifested by a reduction in the water holding capacity (WHC) of flesh of stiffened fish. After resolution of rigor mortis, WHC again rises, though never to the original level. Some other consequences of pH fall are given below.

1. The electric resistance of fish muscle decreases due to increase in the permeability of muscle cells to free ions.

2. The number of free –SH groups increases probably due to the denaturation of proteins following the termination of rigor and the onset of spoilage or both together.
3. The activity of two important enzymes, creatine kinase and ATPase is enhanced at the low pH. These enzymes continuously break down CP and ATP, the two energy-rich organic phosphates. The disappearance of ATP from fish muscle leads to muscle contraction and rigor mortis.

VII. Spread of Rigor Mortis in Fish Body

Like in mammals, in fish also, rigor usually starts at the head and spreads via the chief body muscle to the caudal area in a descending order. That is why, in most species, the mouth and gill cover (operculum) open wide at the beginning of rigor. The resolution of rigor also follows a descending order, spreading from head to tail. However, in some species, the onset of rigor takes place simultaneously over the entire body. The difference in the time of onset of rigor in different body parts is due to the difference in their pH. Rigor may also start at the tail and spreads towards the head. This may occur when fish struggles too much before death with profuse beating of tail resulting in production and accumulation of lactic acid in the tail and quick disappearance of ATP from it.

VIII. Duration of Rigor Mortis

Rigor mortis in fish is of shorter duration than in mammals. The duration varies over a wide range (Table 3).

Table 3: Duration of rigor mortis in fish

	*Marine Fish**	*Freshwater Fish***
(1) Lag of initiation (pre-rigor period)	1–7 hrs	24 hrs
(2) Peak of rigor	5–22 hrs after death	—
(3) Total duration	31–120 hrs	72–80 hrs

*: Marine fish killed by slaughtering and stored in ice.

**: Freshwater fish stored in ice.

IX. Influence of Rigor Mortis on Quality of Fish

Rigor mortis has remarkable influence on the quality of fresh and processed fish as described below.

1. Fresh Fish

Rigor mortis has beneficial effects on the keeping quality of fresh fish. It extends the shelf life of fresh fish by delaying the activity of spoilage microbes, as the low pH produced during rigor is unfavourable for most of them. Hence processes, which delay onset of rigor and prolongs its duration should be undertaken. Such processes include reducing the struggling of fish when caught by gear (*e.g.*, by reducing the trawling period), killing fish by electroshock or by slaughtering immediately after catch, storing fish in chilled condition after harvesting and careful handling of fish.

2. Frozen Whole Fish

Rigor mortis can influence the quality of frozen whole fish in three ways as given below.

(*a*) Gaping

Gaping is the phenomenon due to which individual flakes (myotomes) of muscle get apart giving the tissue a broken and ragged appearance. It occurs when the connective tissues (myocommata) holding together the myotomes (Figure 10) are broken down. Rigor mortis is one of the several reasons of gaping. The skeleton and the connective tissue prevent the contraction of muscle during rigor mortis, but when the contraction force becomes very high, the connective tissue breaks resulting in separation of the myotomes and gaping of the muscle. When such fish is frozen whole and fillets are cut from it after thawing, gaped fillets are obtained. This phenomenon is called fillet gaping. Another reason of gaping is the autolytic hydrolysis of the connective tissue, which is made up of collagen. Collagen of the connective tissue, which holds together the myotomes, is hydrolysed by the enzyme collagenase during post mortem autolysis resulting in gaping. Gaping also occurs due to physical stress on connective tissue due to rough handling. For example, if fish is exposed to high pressure, say by high load of fish and ice above, then also the connective tissue breaks and results in gaping.

(*b*) Toughness of Tissue

The higher the temperature at which fish goes into rigor, the tougher and the more fibrous is the frozen fish when eaten.

(*c*) Drip Loss

The higher the temperature at which fish goes into rigor, the greater is the drip loss. Drip loss is the loss of tissue fluid from the body of frozen fish during thawing. It not only results in weight-loss and loss of nutrients and flavour compounds in the drip, it also affects the eating quality by reducing the pleasant succulence and juiciness of fish tissue.

3. Frozen Fillets

Fillets are the pieces of fish flesh obtained by cutting a fish longitudinally along both the sides of the backbone. Thus each fish gives two fillets. Rigor mortis influences the quality of frozen fillets in the following ways.

(*a*) Shrinkage

In whole fish, the skeleton prevents contraction during rigor mortis. But in fillets, where there is no skeleton, rigor mortis leads to shrinkage of the muscle.

(*b*) Toughness of Tissue

The higher the temperature at which a fillet goes into rigor, the tougher and the more fibrous is the frozen fillet when eaten.

(*c*) Drip Loss

The higher the temperature at which a fillet goes into rigor, the greater is the drip loss. It not only results in weight-loss and loss of nutrients and flavour compounds in the drip, it also affects the eating quality by reducing the pleasant succulence and juiciness of fish tissue.

4. Smoked Fish

When fish is smoked pre-rigor, the treatment in the smoking kiln can cause high temperature rigor effects resulting in very shrunken smoked fillets.

X. Manipulation of Rigor Mortis

Rigor mortis, through a lowered pH, delays the onset of microbial spoilage and tries to keep the freshness of fish for a certain length of time. Therefore, it is always advisable to act in such a way that, the onset of rigor is delayed, its duration prolonged and its magnitude pronounced. Some of the actions to be taken to extend the shelf life of fresh fish through prolonging its duration are given below.

1. Fish should be removed from the gear as quickly as possible and should be killed immediately by slaughtering or electroshock so that it does not struggle too much and does not die of asphyxia.
2. Fish should be handled properly without throwing, bruising or pressing.
3. Fish should be stored at low temperature, say by icing, as quickly as possible.

Fish in the state of rigor is not unfit for use. The later rigor mortis begins and the longer it lasts, the longer fish can be safely stored as rigor mortis delays the onset of subsequent autolytic and microbial spoilage. When fish has to be processed by freezing or smoking, it is always advisable to allow it to complete the rigor process and then to take the post-rigor fish for processing.

XI. Summary

1. What is Rigor Mortis ?

After a short lapse of time from the death of fish, its body remains stiffened for a certain period during which the elastic nature of living muscle disappears and the muscle becomes hard. Such stiffening of muscle after death is called rigor mortis. Prolongation of rigor mortis delays the onset of subsequent deteriorative autolysis and microbial putrefaction thereby delaying the onset of spoilage and retaining the freshness of fish for a longer duration. As a result, the shelf life of fish becomes longer. Therefore, all processing steps undertaken immediately after death of fish aim at the prolongation of the duration of rigor mortis. Depending upon the state of rigor mortis, fish to be utilised for processing can be in any one of the three states such as a) Pre-rigor, from death till the onset of rigor mortis; b) On-rigor, from the onset of rigor mortis till its disappearance and c) Post-rigor, after the disappearance of rigor mortis. When fish is frozen without completing the rigor mortis (pre-rigor or on-rigor), the freezing process arrests the rigor process and the rigor is completed during thawing. This rigor is called 'thaw rigor'. When fish is killed by electric shock, rigor sets in long time after death. Such a rigor is called 'cataleptic rigor'.

2. Stages of Rigor Mortis

Rigor mortis passes through two stages as given below (Figure 2).

(*a*) Stiffening of Muscle

During the stiffening stage, the muscle stiffens gradually resulting in a gradual decrease in the softness from the original softness level. Stiffening continues to a point of maximum stiffness (or lowest softness) called 'peak'. After this starts the second stage of rigor mortis, the softening of muscle.

(*b*) Softening of Muscle

During this stage, the muscle loses its stiffness gradually and softens resulting in a gradual increase in its softness. Softening continues till the softness returns back to original softness level. Rigor mortis is said to be completely resolved or terminated when the softness of fish tissue reaches the original softness level. However, the gradual softening of the muscle continues unabated even after the termination of rigor mortis.

3. Biochemistry of Rigor Mortis

The most important event occurring after the death of fish, which is responsible for rigor mortis, is the break down of its tissue components starting immediately after death. Five of the tissue components, which are most important with respect to post mortem changes are carbohydrates, nucleotides, proteins, lipids (fats) and nucleic acids. Among them, the break down of carbohydrates and nucleotides start immediately after death leading to rigor mortis while the break down of proteins, lipids and nucleic acids takes place afterwards during autolysis, which follows rigor mortis. The five post mortem hydrolytic processes are given below.

<table>
<tr><td>A</td><td>Saccharolysis</td><td>Break down of carbohydrates (glucose and glycogen)</td><td rowspan="2">Start immediately after death</td></tr>
<tr><td>B</td><td>Nucleotidolysis</td><td>Break down of nucleotides (ATP)</td></tr>
<tr><td>C</td><td>Proteolysis</td><td>Break down of proteins</td><td rowspan="3">Start with onset of autolysis which follows rigor mortis</td></tr>
<tr><td>D</td><td>Lipolysis</td><td>Break down of lipids (fats)</td></tr>
<tr><td>E</td><td>Nucleolysis</td><td>Break down of nucleic acids (DNA and RNA)</td></tr>
</table>

(*a*) Saccharolysis

Saccharolysis (*saccharo*: sugar; *lysis*: break down) is the process of breaking down the carbohydrates or sugar substances by a group of carbohydrate-breaking or saccharolytic enzymes, collectively called 'carbohydrases'. The most important carbohydrates present in fish body are glycogen and glucose. The two most important saccharolytic processes taking place in fish body after death are glycogenolysis and glycolysis.

(i) Glycogenolysis

In glycogenolysis, glycogen is hydrolysed to glucose. It can occur in two paths; i) amylolytic path or ii) glycolytic path. The amylolytic path is the principal path in dead fish (Figure 3).

(ii) Glycolysis

In this process, glucose is hydrolysed to produce energy. It can occur in two paths; (1) aerobic glycolysis or (2) anaerobic glycolysis. Out of the two paths, the anaerobic glycolysis path is followed after death as the tissues are no longer supplied with oxygen and anaerobic condition prevails in the tissues. In this path, each glucose is anaerobically converted to two lactic acids with production of 2 ATP (Figure 3).

Thus, after death of fish, glycogen is hydrolysed to glucose by glycogenolysis through the amylolytic path and subsequently this glucose undergoes anaerobic glycolysis with production of two lactic acids and two ATP from each glucose. The lactic acid thus produced accumulates in the fish tissue and lowers its pH. This low pH retards the growth of most of the microbes, thereby hindering the microbial spoilage of fish. This is the reason why rigor mortis, which runs concomitant with acidification of muscle, retards spoilage of fish for a certain length of time and is beneficial for retaining the freshness of fish.

(*b*) Nucleotidolysis

Nucleotidolysis (*nucleotido*: of nucleotide; *lysis*: break down) or nucleotide degradation is the process of break down of nucleotides by a group of enzymes, collectively called 'nucleotidases'. Among the nucleotides, the most important with respect to rigor mortis is adenosine triphosphate (ATP). When fish is alive, it is used as the 'energy currency' of the cells of fish tissue (Figure 4). It is an energy-rich compound, which hydrolyses to liberate energy to be utilised for various metabolic activities of different tissues.

After death of fish degradation of ATP continues irreversibly in a sequence of steps (Figure 5). (1) ATP undergoes hydrolytic dephosphorylation to produce adenosine diphosphate (ADP) and phosphoric acid (P_i) with liberation of large amount of energy. (2) ADP undergoes further hydrolytic dephosphorylation to produce adenosine monophosphate (AMP or adenylic acid) and P_i. (3) AMP undergoes hydrolytic deamination to produce inosine monophosphate (IMP or inosinic acid) and ammonia (NH_3). (4) IMP undergoes hydrolytic dephosphorylation to produce inosine (HxR) and P_i. (5) Inosine is hydrolysed to hypoxanthine (Hx) and D-ribose. (6) Hypoxanthine is sometimes converted to uric acid by the enzyme xanthine oxidase.

It has been found that the ratio of the concentration of hypoxanthine and inosine (instead of hypoxanthine alone) to that of the total ATP and ATP degradation products (all expressed in moles) indicates the extent of deterioration and freshness of fish. This ratio when expressed as percentage is called K value (Text box 1). Thus, K value is an index of freshness of fish in the early stage of deterioration when microbial putrefaction has not started noticeably. 'K value can be defined as the percentage of inosine and hypoxanthine in total ATP and its degradation products present in fish tissue after death'.

$$\text{K value} = \frac{\text{Inosine + Hypoxanthine}}{\text{ATP and its degradation products}} \times 100 = \frac{\text{HxR + Hx}}{\text{ATP + ADP + AMP + IMP + HxR + Hx}} \times 100$$

where, concentrations of all the items are expressed in moles, HxR= Inosine and Hx= Hypoxanthine.

Mechanism of Muscle Contraction During Rigor Mortis

The stiffening of muscle during rigor mortis is basically due to the contraction of traverse skeletal muscle. The theories proposed to explain the mechanism of muscle contraction during rigor mortis are (*a*) Riesser theory, (*b*) Fluckiger theory, (*c*) Embeden theory, (*d*) Szent-Gyorgyi theory and (*e*) Modern theory. The most detailed study on the mechanism of muscle contraction during rigor mortis was carried out by Szent-Gyorgyi (1945). His findings are largely still valid although certain additions have been made to describe it in details through the modern theory. According to the modern theory, it is the disappearance of ATP from the muscle after death, which is responsible for the muscle contraction; not the fall in pH resulting from the production and accumulation of lactic acid formed by the anaerobic break down of tissue glycogen and glucose after death. The modern theory is also called the 'sliding filament-cross bridge theory'. According to this theory, due to the disappearance of ATP, the actin filaments form cross bridges (bonds) with myosin filaments after which the actin filaments slide past the myosin filaments towards the center of the sarcomere resulting in the contraction of the muscle fibers, which leads to stiffening of the muscle during rigor mortis.

When a fish dies, its neuroendocrine system, which controls all the activities going on inside the body of live fish, ceases to function. Due to this, respiration stops leading to decrease in the oxygen content of blood. Simultaneously heart beat stops and blood is no more circulated through the tissues. The tissues no longer get oxygen due to which anaerobic condition prevails after death. But most of the tissues remain alive for sometime after death where metabolic processes continue. Therefore, these tissues need energy, which is obtained from the hydrolysis of ATP. The hydrolysis of ATP leads to a fall in the ATP concentration in the tissues. But the tissues try to maintain the concentration of ATP at a constant level. For a certain lapse of time, fish tissues can maintain the constant ATP level in two ways; (*a*) hydrolysis of phosphocreatine and (*b*) anaerobic oxidation of glucose. Hydrolysis of phosphocreatine continues till its concentration reaches 60 per cent of the original concentration, below which it does not hydrolyse to replenish ATP. Similarly, anaerobic oxidation of glucose to lactic acid stops upon reaching the limiting low pH and it no longer provides energy for replenishing ATP. Therefore, after this period during which ATP break down is reversible, irreversible break down of ATP starts. In this irreversible process, small amount of energy is liberated which is utilised for the residual metabolism. The irreversible break down of ATP leads ultimately to the formation of hypoxanthine and D-ribose.

In the muscle cells of fish (Figure 8), the two lateral tubules of each T system (triad system), which are the transverse tubules of the sarcoplasmic reticulum, maintain a higher concentration of Ca^{++} ions within them as compared to the low concentration of Ca^{++} ions in the sarcoplasm and the myofibrils. The membrane of the lateral tubules has Ca^{++} pumps to pump Ca^{++} ions from the sarcoplasm and myofibrils into the tubules so that a higher concentration of Ca^{++} ions is maintained inside the tubules. The Ca^{++} pumps get energy from ATP and pump Ca^{++} ions uphill against the concentration gradient from the sarcoplasm and myofibrils into the tubules. After the death of fish, ATP gradually disappears due to its irreversible break down. Therefore, the Ca^{++} pumps present on the membrane of the lateral tubules of the T systems do not get energy and cease to function. Now, Ca^{++} ions diffuse freely through the membrane of the lateral tubules downhill along the concentration gradient from the high concentration side within the tubules to the low concentration side in the sarcoplasm and myofibrils (Figure 9). These Ca^{++} ions passing into the sarcoplasm and reaching the myofibrils have a two fold effect which leads to muscle contraction. The effects are given below.

(a) Formation of Actin-Myosin Cross Bridge

When the concentration of Ca^{++} ions in the sarcoplasm exceeds a certain critical level, they bind to the TnC subunit of troponin resulting in a change in the three-dimensional conformation (quaternary structure) of the globular protein troponin (Figures 10 and 11). Due to this, the linkage among TnC, TnI and TnT gets tightened and TnI gets delinked from the actin filament. The tropomyosin filament, which is bonded to TnT subunit, moves deeper into the actin groove exposing the myosin-binding site on the actin filament. Now, the myosin head projecting from the myosin filament towards the actin filament, binds to the exposed myosin-binding site on the actin filament. The bond fonned between the actin and myosin filaments is called 'actin-myosin cross bridge'. The formation of actin-myosin cross bridge results in the formation of actomyosin complex.

(b) Production of Energy for Sliding of Actin Filament Over Myosin Filament

The myosin head has Ca^{++} induced ATPase activity. After death of fish, Ca^{++} ions diffuse freely from the lateral tubules of the T systems and reach the myosin heads. In presence of these Ca^{++} ions, the Ca^{++} induced ATPase of the myosin heads hydrolyses ATP with liberation of energy. This energy is used to bend the series of myosin heads projecting from the myosin filaments towards the center of the sarcomere. As the myosin heads are bonded to the actin filaments, the actin filaments are pulled towards the center of the sarcomere (Figure 12). Once the energy is exhausted, the myosin heads come back to their vertical position and attach to the next binding sites on the actin filaments. The myosin heads again hydrolyse ATP and utilise the energy for bending themselves so as to slide the actin filaments towards the center of the sarcomere. As actin filaments of a sarcomere move closer to each other from both the sides, the two Z lines at the ends of the sarcomere, which provide anchorage to the actin filaments, come closer to each other, thereby reducing the sarcomere length. As the sarcomere length is reduced, the length of the muscle fibers decreases and muscle contracts. Due to contraction of muscle fish body stiffens. In this way, the series of myosin heads act like levers and make oar-like movements to slide the actin filaments over the myosin filaments. This is called 'sliding filament mechanism' of muscle contraction. The muscle remains in the contracted state permanently as Ca^{++} ions are not pumped back to the lateral tubules of the T systems. This permanently contracted condition of muscle after death is called rigor mortis.

Mechanism of Softening of Muscle During Resolution of Rigor Mortis

During rigor mortis, the fish body remains stiffened. After sometime, resolution of rigor mortis starts and the fish muscle begins to soften. Rigor mortis is said to be completely terminated when the softness of fish muscle comes back to the original softness level (Figure 2). Although contraction of muscle during rigor mortis takes place due to formation of bonds between actin and myosin, in softening of muscle during resolution of rigor mortis, it is not the bonds between actin and myosin which are broken, rather the Z lines which provide anchorage to the actin filaments are disintegrated due to their hydrolysis by the endogenous enzymes of fish muscle. The Z lines are made of the protein α-actinin. Towards the end of rigor mortis, autolysis sets in and the endogenous enzymes start hydrolysing the tissue components. When the protein, α-actinin of the Z-lines, is hydrolysed, the actin filaments lose their anchorage, which results in the softening of muscle during resolution of rigor mortis.

4. Measurement of Rigor Mortis

Rigor mortis in fish is measured by determining the 'toughness' of the muscle. Among several methods, the direct method of assessing rigor mortis by measuring toughness of whole fish with the help of an instrument developed by Messtorff (1954) is most important (Figure 14).

5. Factors Affecting Rigor Mortis

If rigor sets in 'early', is of 'shorter duration' and 'lower intensity', it is 'poor rigor' whereas if rigor sets in 'late', is of 'longer duration' and 'higher intensity' it is 'good rigor'. Good rigor is favorable in fish processing point of view as it helps in retaining the freshness of fish for a longer duration. The duration of pre-rigor, that of rigor mortis and its intensity are influenced by intrinsic factors such as species of fish (type of fish, body shape, activeness), its health, feeding condition, age and size as well as by extrinsic factors such as manner of death (out of asphyxia, by hitting on the head of fish, by beheading, by electroshock), handling and storage temperature.

6. Changes During Rigor Mortis

The two most important changes manifested during rigor mortis are a) Loss of texture of fish as shown by pressure marks left when a post-rigor fish is gently squeezed and b) Fall in pH of fish tissue. Immediately after death anaerobic break down of the glycogen to lactic acid starts in the muscle tissue, which continues during rigor mortis. Production and accumulation lactic acid results in a fall in pH of fish tissue from near neutral (7.05-7.35) to about 6.2–6.5 (Figure 48 and Table 2). Fall in pH suppresses the development of spoilage microbes and thereby suppresses the spoilage of fish.

7. Spread of Rigor Mortis in the Body

Like in mammals, in fish also, rigor usually starts at the head and spreads via the chief body muscle to the caudal area in descending order. The resolution of rigor also follows a descending order spreading from head to tail. However, in some species, the onset of rigor takes place simultaneously over the entire body. Rigor may also start at the tail and spreads towards the head. This may particularly occur when fish struggles too much before death with profuse beating of tail resulting in quick disappearance of ATP from it.

8. Duration of Rigor Mortis

Generally fish has the following range of rigor mortis.

	*Marine Fish**	*Freshwater Fish***
(1) Lag of initiation (pre-rigor period)	1–7 hrs	24 hrs
(2) Peak of rigor	5–22 hrs after death	—
(3) Total duration	31–120 hrs	72–80 hrs

*: Marine fish killed by slaughtering and stored in ice.

**: Freshwater fish stored in ice.

9. Influence of Rigor Mortis on Quality of Fish

Rigor mortis has great influence on the quality of fresh and processed fish. Rigor mortis has beneficial effects on the keeping quality of fresh fish. It extends the shelf life of fresh fish by delaying the activity of spoilage microbes, as the low pH produced during rigor is unfavourable for most of them. In frozen whole fish, it can influence the quality through gaping, toughness of tissue and drip loss. In frozen fillets, it also influences the quality adversely through shrinkage, toughness of tissue and drip loss. It influences the quality of smoked fish in that, when fish is smoked pre-rigor, the treatment in the smoking kiln can cause high temperature rigor effects resulting in very shrunken smoked fillets.

10. Manipulation of Rigor Mortis

Rigor mortis, through a lowered pH, delays the onset of microbial spoilage and tries to keep the freshness of fish for a certain length of time. Therefore, it is always advisable to act in such a way that, the onset of rigor is delayed, its duration prolonged and its magnitude pronounced. Some of the actions to be taken to extend the shelf life of fresh fish through prolonging its duration are (1) Fish should be removed from the gear as quickly as possible and should be killed immediately by slaughtering or electroshock so that it does not struggle too much and does not die of asphyxia, (2) Fish should be handled properly without throwing, bruising or pressing and (3) Fish should be stored at low temperature, say by icing, as quickly as possible.

Fish in the state of rigor is not unfit for use. The later rigor mortis begins and the longer it lasts, the longer fish can be safely stored as rigor mortis delays the onset of subsequent autolytic and microbial spoilage. When fish has to be processed by freezing or smoking, it is always advisable to allow it to complete the rigor process and then to take the post-rigor fish for processing.

4

Autolysis

I. What is Autolysis ?

Autolysis (*auto*: self; *lysis*: break down) is the process of break down of fish tissue after death by its own body enzymes. It manifests itself mainly through softening of the tissue. The complex tissue components are broken down or hydrolysed into their simpler building units by the native or endogenous enzymes[1] present in the fish tissue. The five most important tissue components broken down during autolysis are proteins, lipids, nucleic acids, carbohydrates and nucleotides.

The break down of two of the above tissue components namely, the carbohydrates (glycogen and glucose) and the nucleotide (ATP) starts immediately after death. Therefore, autolysis, in strict sense, starts immediately after death. However, the most commonly referred autolysis, which manifests itself in the gradual softening of the muscle above the original softness level (Figure 2) starts only after rigor mortis is terminated. The low pH attained from break down of glycogen to lactic acid after death enhances the activity of the endogenous tissue enzymes, which can break down the remaining three tissue components such as proteins, lipids and nucleic acids. During autolysis, which follows rigor mortis, these enzymes break down the proteins, lipids and nucleic acids into their simpler building units.

Again, among the three components, which are hydrolysed during autolysis, hydrolysis of proteins is of great significance. They are the main tissue components of fish constituting about 80 per cent on dry weight basis. The muscle proteins such as actin, myosin and collagen are the main structural components of fish tissue. In live condition, these proteins remain in their natural state supporting the original tissue structure and maintain the original softness level. The low pH obtained during rigor mortis enhances the activity of the proteolytic (protein-breaking) enzymes, particularly that of the cathepsins, which are the main proteolytic enzymes of fish muscle. These proteolytic enzymes degrade the proteinous structural tissue components of fish through the process called 'proteolysis' resulting in softening of the tissue. As autolysis manifests itself mainly through softening of tissue which is caused by proteolysis, often proteolysis is regarded same as autolysis.

---

1 Enzymes, which can hydrolyse tissue components of fish, are of two types as follows.

1. **Endogenous or native enzymes:** They are synthesised by the fish body and stored inside it for use in the metabolic processes as and when required. They include the enzymes present in the muscles, gut, pancreas, spleen, liver, kidney etc.

2. **Exogenous or foreign enzymes:** They are not synthesised by fish body. They may be synthesised inside the fish body and available there, say the microbial enzymes synthesised by the natural microbial flora of fish and available there to hydrolyse the fish tissue components after death or may be enzyme preparations added to the minced fish to break down fish tissue quickly, say during the preparation of fish hydrolysate.

II. Types of Enzymes Involved in Autolysis

Endogenous enzymes are the agents of autolysis. Live fish contains numerous enzymes required for the complex chemical changes, collectively called metabolism, going on inside the fish body. These enzymes are distributed throughout the body. Out of the several functions carried out by them in different ways, one function is to hydrolyse (break down or degrade) complex substances into simpler forms. This group of enzymes, which is involved in the hydrolysis process, is called hydrolases[2]. It is only this group of enzymes, which is responsible for the autolytic break down of tissue components. When a fish is alive, these enzymes have controlled or regulated activity. They are put to use as and when required and prevented from their action in rest of the time.

With respect to the location in fish body and the mode of regulation of their activities, the hydrolases involved in autolysis are of two types.

A. Digestive or gut enzymes and

B. Muscle and internal organ enzymes.

A. Digestive or Gut Enzymes

The digestive or gut enzymes are those enzymes, which are secreted into and/or are present in the gut to digest the food taken by fish (Figure 15). These enzymes are synthesised within the secretory cells of the digestive system (*e.g.* those of pancreas, intestine etc.) and are stored within these cells as inactive precursors called 'proenzymes' or 'zymogens'. When food enters into the gut, the neuroendocrine system of fish sends information to the secretory cells to secret the zymogens to outside. Thus the enzymes of the gut are 'extracellular enzymes'. Outside the cell, they are converted to active enzymes in several ways such as by lowering of pH or by action of some other enzymes. Now, the active enzymes act on the complex food components and digest them into simpler forms, which are absorbed into the wall of the intestine for further assimilation. Thus, in living fish the hydrolases do not act on the own tissue due to their presence as inactive zymogens. However, their presence as inactive zymogens or active enzymes is controlled by the neuroendocrine system. The zymogens of the digestive proteolytic enzymes pepsin, trypsin and chymotrypsin are pepsinogen, trypsinogen and chymotrypsinogen respectively.

After death of fish, some enzymes, which have already been secreted into the gut to digest food when fish is alive, remain inside it as residual enzymes. Besides, the neuroendocrine system, which controls the activity of the enzymes, ceases to function due to which more of zymogens are secreted and converted to active enzymes. These freshly secreted enzymes along with the residual enzymes accumulate inside the gut. These gut enzymes, which are mostly proteolytic enzymes such as pepsin, trypsin and chymotrypsin, hydrolyse the proteins of the gut wall and enter into the muscle. There they hydrolyse the muscle proteins due to which the muscle adjacent to the gut becomes very soft and flabby. As the enzymes move gradually from the gut towards the body surface through the muscle, the muscle degradation and softening proceeds from the gut towards the skin.

---

2 Hydrolases is the third class of enzyme based on the International Classification of Enzymes as shown below.

	Class	*Type of Reaction Catalysed*
1.	Oxidoreductases	Transfer of electrons
2.	Transferases	Group transfer reactions
3.	**Hydrolases**	**Hydrolysis reactions (transfer of functional groups to water)**
4.	Lyases	Addition of groups to double bonds or the reverse
5.	Isomerases	Transfer of groups within molecules to yield isomeric forms
6.	Ligases	Formation of C–C, C–S, C–O and C–N bonds by condensation reactions coupled to ATP cleavage

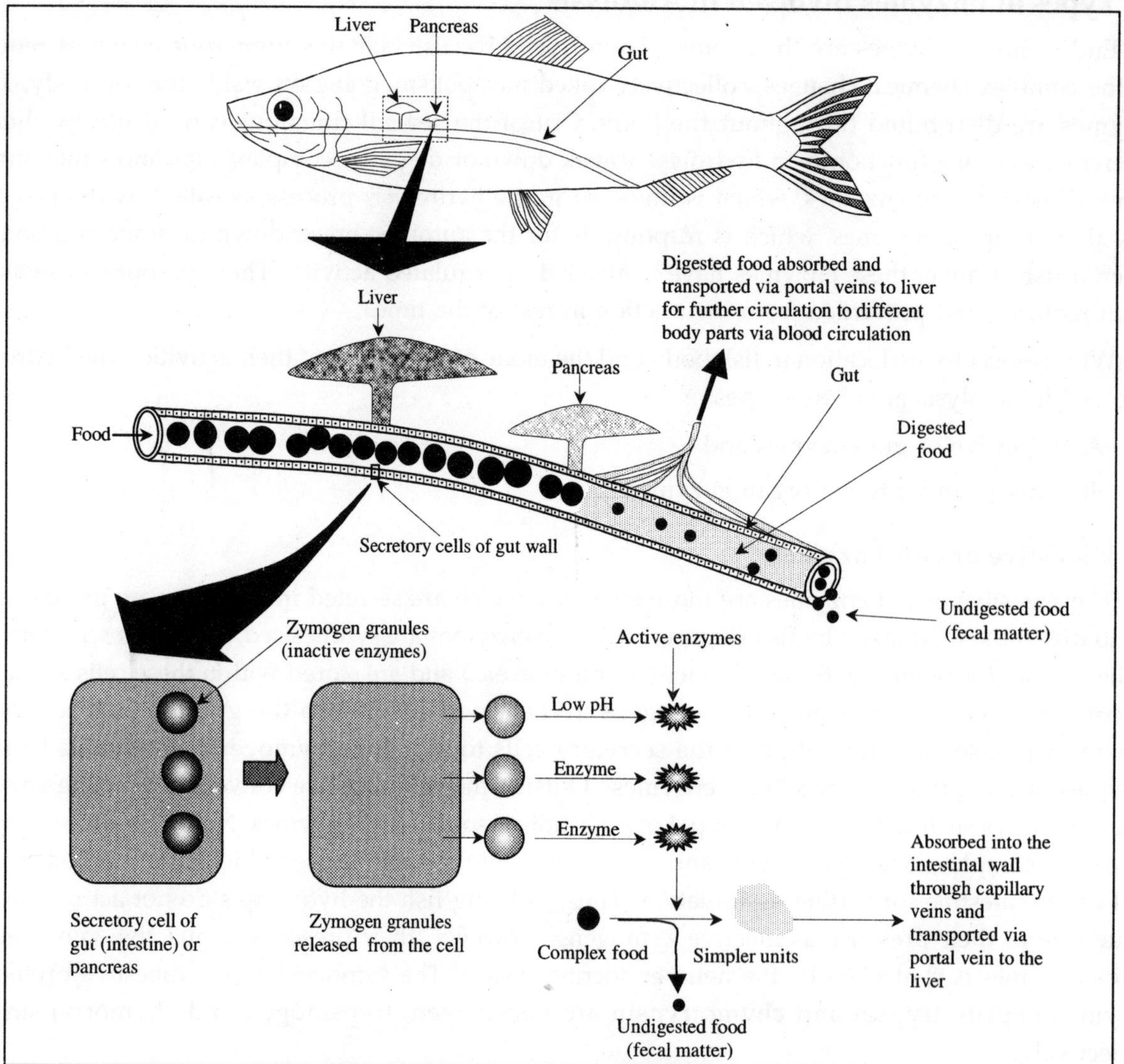

Figure 15: Activities of digestive enzymes in living fish

In live fish, the gut has a natural microbial flora, which cannot enter through the gut wall. The hydrolysis of the gut wall after death allows this microbial flora to penetrate through it into the muscle. The hydrolysis of the proteins of the gut wall and muscle surrounding the gut into free amino acids produces a nutrient-rich simple medium for the growth of this microbial flora. They grow profusely utilising the free amino acids and degrade them to off-odour, off-flavour substances presence of which distracts the consumers thereby resulting in the spoilage of fish. Therefore, it is advisable to remove the gut of fish immediately after harvesting, whenever possible, to prevent early spoilage. However, when the catch is very high or when the size of fish is too small to remove the gut, gutting is not practicable commercially.

B. Muscle and Internal Organ Enzymes

The hydrolytic enzymes of the muscle and internal organs such as liver, kidney, spleen and gonads are intracellular enzymes and are present inside the cells, confined within minute membrane-bound organelles present in the cell cytoplasm. These cytoplasmic enzyme-rich minute organelles are called lysosomes (*lysis*: dissolution, *some*: body). They are present in all eukaryotic cells except in erythrocytes. In the muscle cells, they are present in the peripheral sarcoplasm (cytoplasm of muscle cell). The activity of the lysosomal enzymes is controlled by the lysosomal membrane, which encloses them and keeps them separated from the cytoplasm and other cytoplasmic organelles. The enzymes, otherwise, would digest them. When required, the membrane ruptures releasing the enzymes to digest the given material intracellularly or extracellularly. In living tissues, the function of lysosome is to digest, degrade and remove dead cells. There are 50 types of hydrolytic enzymes (hydrolases) present in the lysosome, which can digest most varieties of tissue components such as proteins, fats and nucleic acids. However, they cannot hydrolyse the lysosomal membrane. In this way, this membrane does not allow the hydrolases to hydrolyse the own live cells. The important lysosomal enzymes are cathepsins (proteolytic enzymes of muscle), lipases, carbohydrases, ribonucleases (RNase) and deoxyribonucleases (DNase). In live tissue these enzymes have three functions as given below.

1. Endocytosis

The cell engulfs a foreign particle into the cytoplasm. The lysosomes release their enzyme secretion on the particle and digest it intracellularly.

2. Exocytosis

The lysosomes move to the cell membrane and rupture upon touching it. The enzymes are released to outside the cell where they digest any particle extracellularly.

3. Autophagy

The lysosomes secret their enzymes on intracellular organelles like mitochondria or golgi complex present within the same cell and digest them. Therefore, lysosome is called the 'suicide bag of the cell'.

After death of fish, when its chemical conditions change, the membranes of the lysosomes get damaged liberating the hydrolytic enzymes into the cytoplasm. They get damaged easily when oxygen supply to the cells stops after death. The low pH produced in the fish muscle due to the production and accumulation of lactic acid from anaerobic oxidation of glucose after death favours most of the hydrolytic enzymes of the lysosomes. Now, these enzymes hydrolyse the cellular components such as proteins (most prominent), carbohydrates, fats and nucleic acids. The enzyme collagenase is also present in the muscle which breaks down the collagenous connective tissue holding the muscle components together and binding them to the bones.

The most important lysosomal enzyme causing autolysis is cathepsins, a group of proteolytic enzymes. It is the most important enzyme responsible for the degradation of proteins of the muscle and internal organs. It is a group of acid proteases. In the muscle cell, cathepsins break down the myofibrillar proteins (actin and myosin) and sarcoplasmic proteins (myogen and globulin) to amino acids. Different types of cathepsins have been identified (cathepsin A, B, C, D, E, H, L etc.) out of which cathepsin A, B, C, D and L are present in fish muscle. Fish muscle contains more cathepsins than muscle of other animals. This is one of the reasons why fish spoils more quickly than other animals. However, the concentration of cathepsins in the muscle is much less than that in the internal organs such as kidney, liver, spleen and gonads. Therefore, immediately after harvesting, these internal organs are removed alongwith gut from the fish body to prevent autolysis. In other words, the viscera (gut + kidney, liver, pancreas, spleen, gonads etc.) are removed immediately after harvesting. A second group of intracellular proteases called calpains or calcium activated factor (CAF) hydrolyse the myosin, troponin and Z line proteins leading to post mortem tenderisation of fish muscle. The third group of intracellular proteases is collagenase, which hydrolyses the protein collagen. As the connective tissue, which is made of collagen, holds together the myotomes (muscle compartments), hydrolysis of collagen by collagenase results in undesirable separation of the myotomes into flakes. This phenomenon is called 'gaping'. Gaping of fillets results in its poor appearance resulting in consumers' dislikingness (p. 27).

III. Autolytic Hydrolysis

The autolytic break down of the important fish tissue components takes place in the following five steps.

1. **Saccharolysis:** Complex carbohydrates —*Carbohydrases*→ Simple sugars
2. **Nucleotidolysis**: Nucleotide (ATP) —*Nucleotidases*→ Phosphates + sugar (D-ribose) + purine (hypoxanthine)
3. **Proteolysis**: Proteins —*Proteases*→ Amino acids
4. **Lipolysis**: Lipids (fats) —*Lipases*→ Free fatty acids + glycerol
5. **Nucleolysis**: Nucleic acids (DNA & RNA) —*Nucleases (DNase/RNase)*→ Nucleotides

Out of the above five steps, saccharolysis and nucleotidolysis take place immediately after death. They have been described under rigor mortis. The process of true autolysis, which follows rigor mortis and which manifests itself in softening of the muscle has the following three steps.

1. Proteolysis
2. Lipolysis
3. Nucleolysis

1. Proteolysis

The tissue proteins of fish are broken down by a group of endogenous protein-breaking or proteolytic (*proteo*: proteins; *lytic*: breaking) enzymes collectively called 'proteases' (Figure 16). Proteins are sequentially broken down by the proteases to proteoses, peptones, polypeptides and finally to amino acids. This process is called 'proteolysis'. The tissue proteins are hydrolysed both by the gut enzymes and the enzymes of the muscle and internal organs. The gut enzymes such as pepsin, trypsin and chymotrypsin hydrolyse the proteins of the gut and that of the muscle adjacent to the gut. On the other hand, cathepsins, which is the most important lysosomal enzyme of the muscle and the internal organs, hydrolyse the proteins of the muscle and internal organs.

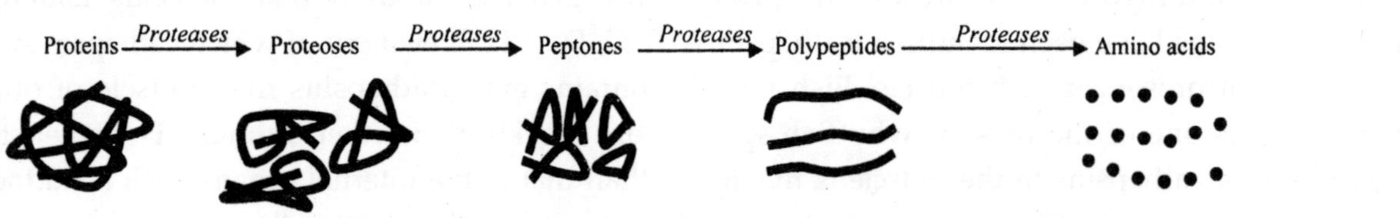

Figure 16: Break down of proteins during proteolysis

2. Lipolysis

The tissue lipids of fish, which consist mainly of the neutral lipids (triglycerides / triacylglycerols / fats) are broken down by a group of lipid-breaking or lipolytic (*lipo*: lipid; *lytic*: breaking) enzymes collectively called 'lipases' (Figure 17). The triglycerides are sequentially degraded to diglycerides, monoglycerides and finally to free fatty acids and glycerol. The gut lipases can hydrolyse the linkage at all the three positions (1st, 2nd and 3rd) unlike many other lipases, which often do not hydrolyse the linkage at the 2nd position. Another important group of lipids called the polar lipids (phospholipids) are broken down by the enzymes 'phospholipases' to glycerol, free fatty acids, phosphoric acid and an alcoholic group (choline, ethanolamine, serine or inositol). Both lipases and phospholipases are found in gut, muscle and internal organs. The free fatty acids (FFA) produced by the process of lipolysis accumulate in the fish tissue and result in the development of a typical rancidity called 'hydrolytic rancidity' (Text box 7, p. 80).

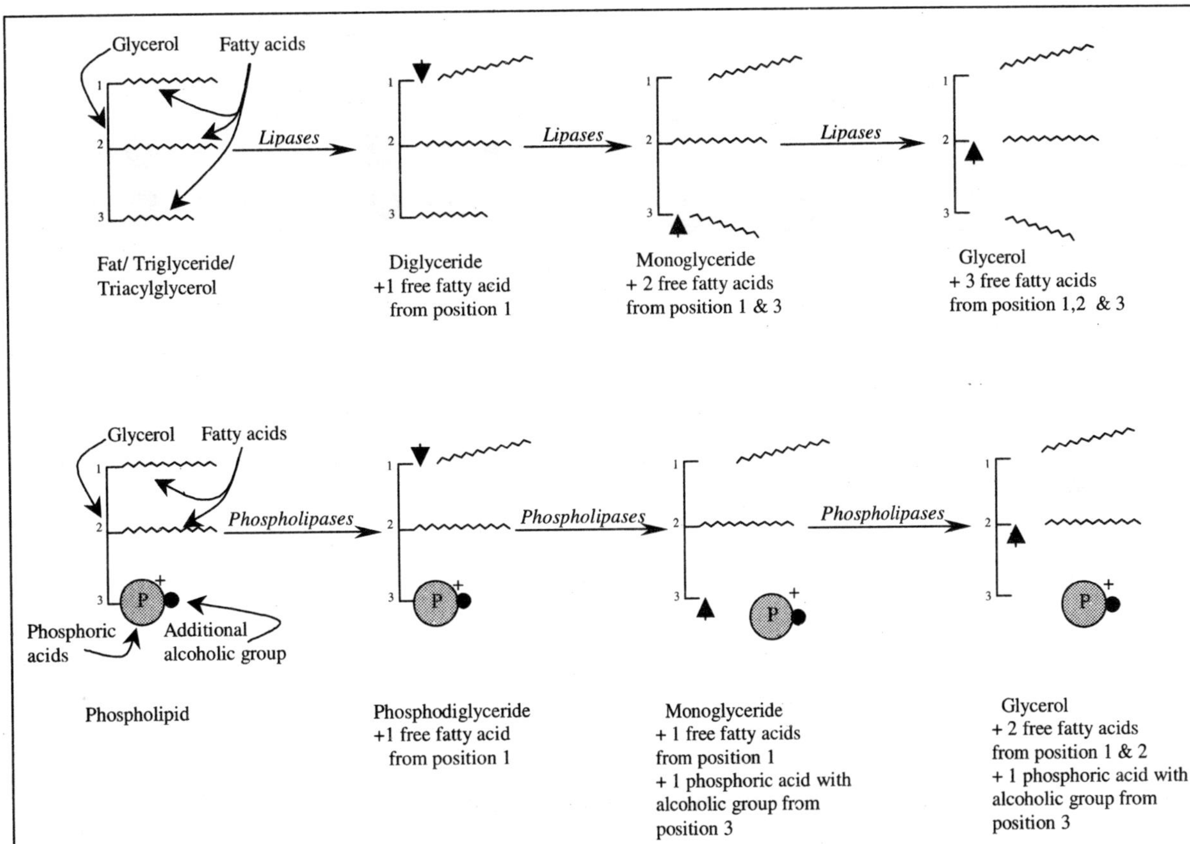

Figure 17: Break down of fats and phospholipids during lipolysis

3. Nucleolysis

The nucleic acids are present in all cells in two forms such as deoxyribonucleic acid (DNA) and ribonucleic acid (RNA). During autolysis, the nucleic acids of fish tissues are hydrolysed by a group of endogenous enzymes collectively called 'nucleases' (Figure 18). DNA is hydrolysed by the enzyme deoxyribonuclease (DNase) and RNA by the enzyme ribonuclease (RNase). They are hydrolysed to nucleotides, which are the building blocks of these nucleic acids. Nucleases are present in the gut, muscle and internal organs of fish.

IV. Effects of Autolysis

Autolysis has both negative and positive effects. In fresh fish, autolysis has more pronounced negative effect on the quality. On the other hand, there are some processing methods, say preparation of fish silage and fish hydrolysate, where autolysis has more pronounced positive effect. However, the best way of fish utilisation is by direct consumption. This needs fish to keep fresh. Autolysis adversely affects freshness. Thus the negative effects of autolysis are of greater concern than the positive effects. By understanding the effects of autolysis, the rate of autolysis can be enhanced or retarded as desired.

A. Negative Effects

1. Softening of Tissue (Loss of Firmness)

The texture of fresh fish is firm (consistent) and elastic. It does not retain finger indentations. If the flesh is pressed by finger, it does not retain the pressure mark of the finger. But autolysis breaks down the proteinous structural components (actin, myosin and collagen), which is manifested by softening of the muscle below the original softness level of fresh fish (Figure 2). The fish becomes soft and the firm elastic texture disappears. Now, the fish retains finger indentations. Texture of fish is most important in organoleptic (sensory) evaluation of quality of fish, in determining the cutting efforts required and in

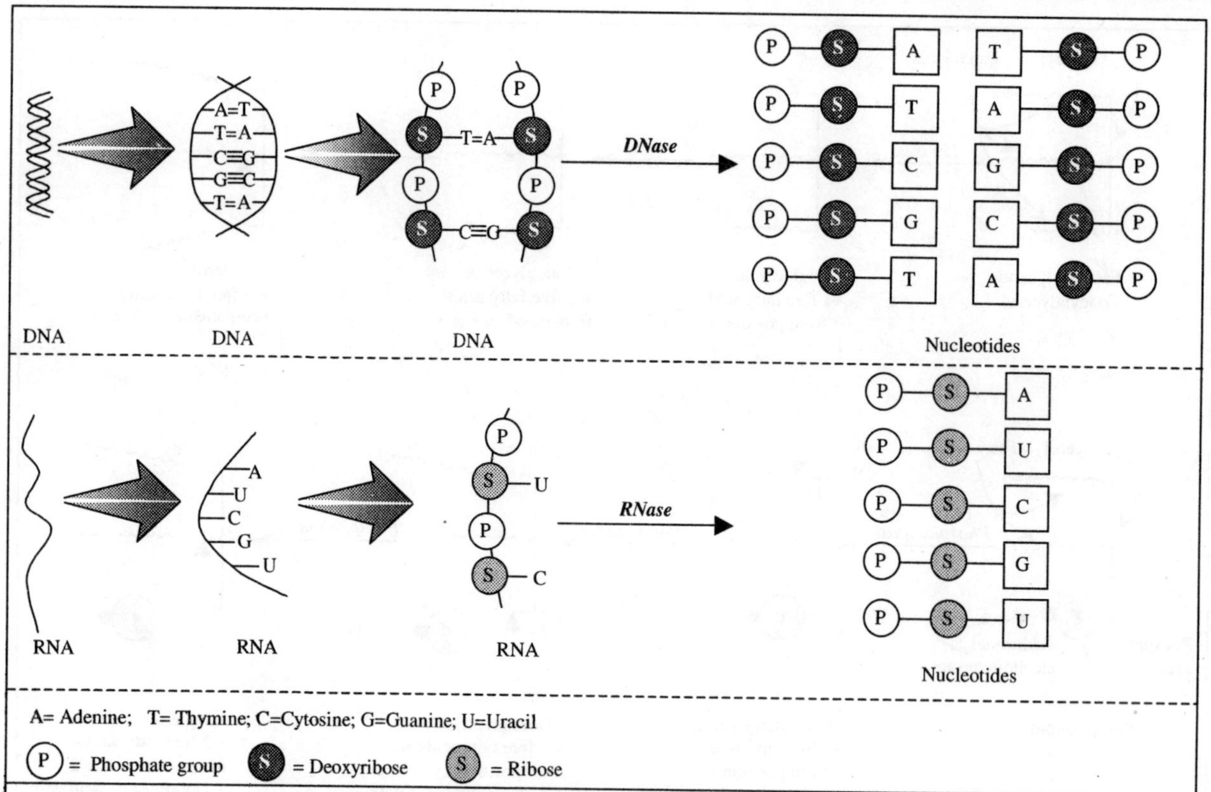

Figure 18: Break down of nucleic acids during nucleolysis

determining the ease of handling fish. As the texture of fish becomes soft due to autolysis, (1) Fish gets low sensory scores and distracts the consumers, (2) It requires more cutting efforts and (3) It is difficult to be handled and processed. For example, dressing and filleting becomes difficult and during frying, it breaks into smaller pieces. In chilled prawn, the relatively short shelf life is due to softening of the tissue by the action of the proteolytic enzyme collagenase present in its hepatopancreas.

2. Penetration of Microbes

In live fish, microbes are present on the surfaces exposed to outside. Skin, gills and lumen of the gut contain high microbial loads in their increasing order (Figure 19). The lumen of gut, though present inside the fish body, is exposed to outside through mouth and cloacal aperture. The muscle and internal organs of fresh fish, being unexposed to outside, are sterile (free from microbes). The natural microbial flora of the gut lumen cannot enter through the gut walls into the muscle in live fish. However, after death, when the digestive enzymes of the gut hydrolyse the gut wall during autolysis, the microbial flora penetrates through it into the hitherto sterile muscle and internal organs. This microbial flora acts on the muscle and internal organs and putrefies them. Similarly, when the components of the skin are hydrolysed by autolysis, it becomes soft and permeable to microbes. The microbial flora present on the skin penetrates through it into the muscle and internal organs and putrefies them.

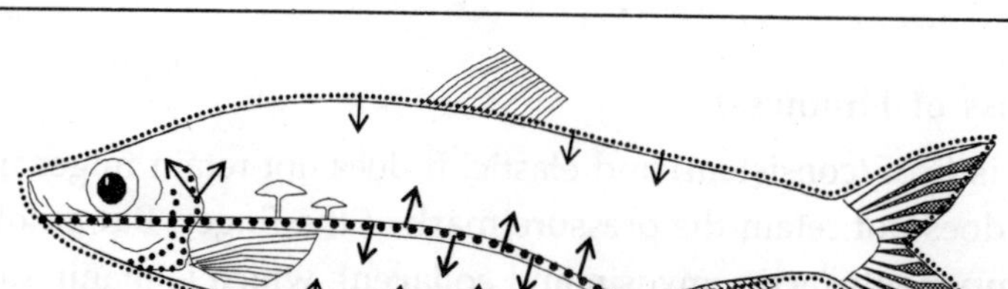

Figure 19: Penetration of microbes into fish muscle after death

3. Growth of Microbes

During autolysis, the complex tissue components are hydrolysed to simpler substances, which become a nutrient-rich medium for microbial growth. The microbes utilise these readily available simpler substances for their growth and metabolise them to off-odour, off-flavour substances presence of which renders the fish spoiled. If autolysis could be arrested (or even slowed down), microbes would have to first secret their own enzymes to the surrounding to hydrolyse the tissue components to simpler substances after which they can absorb and utilise them (Figure 45). This process is difficult and time taking due to which spoilage could be delayed.

4. Increase in NPN Content

The non-protein nitrogenous (NPN) substances such as TMAO, urea and amino acids are the most favoured substances for microbial growth. Small quantities of NPN are present in live fish, but during autolysis, due to hydrolysis of proteins, substantial quantity of free amino acids is produced (Figure 15). This increases the NPN in fish tissue making it a nutrient-rich medium for microbial growth.

5. Development of Rancidity

The hydrolysis of fish lipids results in the formation of free fatty acids (FFA) as one of the end products (Figure 17). Its accumulation in fish tissue results in the development of a type of rancidity called 'hydrolytic rancidity'.

6. Belly Bursting

It is a common post harvest problem in some species of pelagic fishes like oil sardine[3], herring[4], capelin[5] and sprats[6]. These species, during a particular period, usually in the post-spawning seasons, show high intake of food. When food enters into the gut, the neuroendocrine system induces the secretory cells of the organs of digestive system (pancreas, intestine) to secret large amount of enzymes in the form of inactive zymogens. The zymogens become active enzymes outside the secretory cells and act on the food components. If these fishes are caught during such periods of high food intake, the large quantities of enzymes secreted into the gut remain active for quite a long period of time after death. They degrade the gut wall completely within few hours and enter into the muscle. First, they hydrolyse the muscle adjacent to the gut. The end products of the hydrolysis of gut and adjacent muscle are simple substances like amino acids, which become a nutrient-rich medium for the activity of the natural microbial flora of the gut. Besides, the gut of fish having high intake of food contains a very high load of microbes, which have entered into the gut alongwith the food. All these microbes grow on the end products of hydrolysis resulting in the production of gases like CO_2 and H_2. The accumulation of these gases inside the belly cavity leads to the swelling of the belly. The belly, on the other hand, becomes soft due to the hydrolysis of its tissues by the gut enzymes. Ultimately the belly bursts after a short period from death. This phenomenon is called 'belly bursting'. Fishes showing belly bursting degrade quickly and therefore spoil easily soon after they are caught.

The problem of belly bursting can be overcome by adopting certain measures. (1) By gutting the fish immediately after harvesting as the gut enzymes are lost alongwith the gut. However, gutting is not possible in case of small and bulk catches. (2) By keeping the fish alive after harvesting so as to allow the ingested food to pass out from the gut as feces thereby reducing the quantity of digestive enzymes in the gut. This method is not practicable in most cases. (3) By quick chilling of fish immediately after harvesting. It is the most effective method of preventing belly bursting.

---

3 *Oil sardine*: *Sardinella longiceps*.

4 *Herring*: Oceania herring (*Aldrichetta forsteri*), Atlantic herring (*Clupea harengus harengus*), Pacific herring (*C. harengus pallasii*).

5 *Capelin*: Arctic capelin (*Mallotus catervarius*), Atlantic capelin (*M. villosus*), Pacific capelin (*M. socialis*).

6 *Sprats*: European sprats (*Clupea sprattus*), New Zealand sprats (*Maugeclupea antipodum*).

7. Gaping

The skeletal muscle forming both the sides of fish body are made of several transverse muscle segments called myotomes, which are separated from each other by connective tissue sheets called myocommata (Figure 10). Within each myotome remain large numbers of elongated muscle cells, which run parallel to each other longitudinally along the length of the fish body each extending between two myocommata. The myocommata is made of collagen fibers. Thus, the whole muscle is supported by the collagenous connective tissue, myocommata. During autolysis, the proteolytic enzyme collagenase breaks down the protein collagen resulting in the disintegration of the myocommata. This results in undesirable separation of the myotomes into flakes. This phenomenon is called gaping. Gaping results in its poor appearance resulting in consumers' dislikingness (p. 27).

The end products of autolysis are free amino acids, glycerol, free fatty acids etc., which do not impart much objectionable off-flavour or off-odour to fish. A fish, which has passed through autolysis, is still acceptable to the consumer. However, the end products of autolysis become a nutrient-rich medium for the microbes. They utilise these simple end products and convert them to off-odour, off-flavour substances responsible for the spoilage of fish. Therefore, autolysis, though not spoilage in itself, leads to easy spoilage by microbes.

B. Positive Effects

1. Taste and Flavour Development

The intermediate products of nucleic acid (DNA and RNA) break down during autolysis and nucleotide (ATP) break down immediately after death include adenosine monophosphate (AMP), inosine monophosphate (IMP) and guanosine monophosphate (GMP). The products of protein hydrolysis during autolysis include amino acids like histidine, glutamic acid, glycine, valine, alanine, arginine and methionine. All these break down products are flavour potentiators due to which they impart pleasant flavour and taste to fish. Thus, some of the products of autolysis are responsible for taste and flavour development in fish.

2. Liquefaction

Fish silage is a byproduct prepared from uneconomic fish and fish offal by allowing the fish results flesh to liquefy completely. The endogenous body enzymes hydrolyse the tissue components, which results in a complete liquefaction of fish flesh. The hydrolysis is enhanced by adding acids, which lower the pH of fish tissue to the point favourable for the activity of the endogenous enzymes.

3. Maturation of Salted Fish

In the temperate Scandinavian countries, salted fish is stored at cooler temperatures for months together (sometimes upto nine months) to undergo the process of 'maturation' or 'ripening'. It is especially important for salting of fishes like herring and anchovy, which have high fat content. During this period, the endogenous tissue enzymes hydrolyse the tissue components to simpler compounds, which impart the typical pleasant mature taste greatly relished by the people of these countries. Fish loses its raw flavour and odour and its flesh becomes tender, which separates readily from the bones. Salted fish prepared in this way is eaten without cooking. Cathepsin is the enzyme responsible for the textural changes (softening) during the ripening of salted fish.

4. Tenderisation

Autolytic hydrolysis of the fish proteins results in the tenderisation of fish meat. Tenderised meat is soft and pliable for which it is organoleptically better accepted than the tough pre-tenderised meat.

V. Measurement of Autolysis

The extent of autolysis undergone after death is measured by estimating the concentration of the end products of autolysis such as free amino acids and free fatty acids. Their concentration is measured in terms of the following parameters.

1. Alpha Amino Nitrogen (AAN)

It is an estimate of the concentration of free amino acids present in the fish tissue. Each amino acid has an amino group at its alpha position. In proteins, the amino acids, which are the building blocks of proteins, are bonded to each other in a chain by peptide bonds. These bonds involve the alpha amino groups of the amino acids. Autolytic hydrolysis of proteins breaks the peptide bonds and makes the amino acids free. Due to this, the alpha amino groups of each free amino acid becomes free. Each alpha amino group contains a nitrogen atom. Thus the concentration of alpha amino nitrogen in fish indicates the extent of autolytic hydrolysis. 'Alpha' amino nitrogen is estimated, because some amino acids also contain amino group in other position. For example, the amino acid lysine contains another amino group in its 'epsilon' position besides that at the alpha position and some amino acids contain nitrogen in positions other than the alpha amino group.

2. Non-Protein Nitrogen (NPN)

When proteins are hydrolysed to amino acids, the free amino acids produced no longer remain part of the proteins and therefore count towards non-protein component. Fish tissue also contains other non-protein nitrogenous substances such as TMAO, creatine, urea etc. Thus the quantity of non-protein nitrogenous substances in fish expressed in terms of non-protein nitrogen (NPN) partly indicates the extent of autolytic hydrolysis of proteins.

3. Free Fatty Acids (FFA)

During autolysis, fat is hydrolysed to free fatty acids and glycerol. The amount of free fatty acids (FFA) in fish tissue indicates the extent of hydrolysis of fats and therefore indicates the extent of autolysis.

VI. Control of Autolysis

The rate of autolytic degradation of tissue components mainly depends on the following factors.

1. Temperature
2. Presence of gut and internal organs
3. Moisture content of tissue
4. Salt content and
5. pH

If it is required to retard autolysis, as in case of fresh fish, which should not be allowed to soften by autolysis so as to retain its freshness, then the following steps should be taken.

1. Evisceration

Evisceration involves removal of the gut and internal organs. This process retards autolysis, as it does not allow the enzymes of the gut and internal organs to remain in the fish body and hydrolyse the tissues.

2. Low Temperature

Temperature of fish should be lowered quickly by chilling or freezing, which retards enzyme activity.

3. High Temperature

Fish should be exposed to very high temperature by cooking, canning or frying, which denatures the tissue enzymes beyond a certain limiting temperature.

4. Drying

Drying retards autolysis by reducing the moisture content of the fish tissue to such a low level where enzyme activity is almost negligible. Enzymes remain active only in their native three-dimensional conformations, which need water to be present in their structures.

5. Salting

Salting also retards autolysis, because the high concentration of salt used in salting preservation of fish denatures most of the tissue enzymes.

On the other hand, autolysis can be enhanced if required, as in case of preparation of fish silage, by the following ways.

1. Lowering of pH

Autolysis is enhanced by lowering the pH through addition of acids to the point, which favours maximum activity of the autolytic tissue enzymes.

2. High Temperature

As enzyme activity increase with increasing temperature upto the limiting temperature of denaturation, autolysis can be accelerated by exposing fish to high temperatures not reaching this limiting temperature.

VII. Summary

1. What is Autolysis ?

Autolysis (*auto*: self; *lysis*: break down) is the process of break down of fish tissue after death by its own body enzymes. It manifests itself mainly through softening of the tissue. The complex tissue components are broken down or hydrolysed into their simpler building units by the native or endogenous enzymes present in the fish tissue. The five most important tissue components broken down during autolysis are proteins, lipids, nucleic acids, carbohydrates and nucleotides. The break down of two of the above tissue components namely, the carbohydrates (glycogen and glucose) and the nucleotide (ATP) starts immediately after death. Therefore, autolysis, in strict sense, starts immediately after death. However, the most commonly referred autolysis, which manifests itself in the gradual softening of the muscle above the original softness level (Figure 2) starts only after rigor mortis is terminated. The low pH attained from the break down of glycogen to lactic acid after death enhances the activity of the endogenous tissue enzymes, which can break down the remaining three tissue components such as proteins, lipids and nucleic acids. During autolysis, which follows rigor mortis, these enzymes break down the proteins, lipids and nucleic acids into their simpler building units.

2. Types of Enzymes Involved in Autolysis

With respect to the location in fish body and the mode of regulation of their activities, the enzymes involved in autolysis are of two types.

A. Digestive or gut enzymes and

B. Muscle and internal organ enzymes.

A. Digestive or Gut Enzymes

The digestive or gut enzymes are those enzymes, which are secreted into and/or are present in the gut to digest the food taken by fish (Figure 15). These are 'extracellular enzymes'. Outside the cell, they are converted to active enzymes. After death of fish, some enzymes, which have already been secreted into the gut, remain inside it as residual enzymes. Besides, the neuroendocrine system, which controls the activity of the enzymes, ceases to function due to which more of zymogens are secreted and converted to active enzymes. These gut enzymes, which are mostly proteolytic enzymes such as pepsin, trypsin and chymotrypsin, hydrolyse the proteins of the gut wall and enter into the muscle. There they hydrolyse the muscle proteins due to which the muscle adjacent to the gut becomes very soft and flabby.

B. Muscle and Internal Organ Enzymes

The hydrolytic enzymes of the muscle and internal organs such as liver, kidney, spleen and gonads are intracellular enzymes and are present inside the cells, confined within minute membrane-bound organelles present in the cell cytoplasm. These cytoplasmic enzyme-rich minute organelles are called lysosomes (*lysis*: dissolution, *some*: body). The most important lysosomal enzyme causing autolysis is cathepsins, a group of proteolytic enzymes. Different types of cathepsins have been identified (cathepsin A, B, C, D, E, H, L etc.) out of which cathepsin A, B, C, D and L are present in fish muscle. The concentration of cathepsins in the muscle is much less than that in the internal organs.

3. Autolytic Hydrolysis

The autolytic break down of the important fish tissue components takes place in the following five steps.

(*a*) **Saccharolysis:** Complex carbohydrates —*Carbohydrases*→ Simple sugars

(*b*) **Nucleotidolysis**: Nucleotide (ATP) —*Nucleotidases*→ Phosphates + sugar (D-ribose) + purine (hypoxanthine)

(*c*) **Proteolysis**: Proteins —*Proteases*→ Amino acids

(*d*) **Lipolysis**: Lipids (fats) —*Lipases*→ Free fatty acids + glycerol

(*e*) **Nucleolysis**: Nucleic acids (DNA & RNA) —*Nucleases (DNase/RNase)*→ Nucleotides

Out of the above five steps, saccharolysis and nucleotidolysis take place immediately after death. They have been described under rigor mortis. The process of true autolysis, which follows rigor mortis and which manifests itself in softening of the muscle has the following three steps.

(*a*) Proteolysis

(*b*) Lipolysis

(*c*) Nucleolysis

(*a*) Proteolysis

The tissue proteins of fish are broken down by a group of endogenous protein-breaking or proteolytic (*proteo*: proteins; *lytic*: breaking) enzymes collectively called 'proteases' (Figure 16). Proteins are sequentially broken down by the proteases to proteoses, peptones, polypeptides and finally to amino acids. This process is called 'proteolysis'.

(*b*) Lipolysis

The tissue lipids of fish, which consist mainly of the neutral lipids (triglycerides / triacylglycerols / fats) are broken down by a group of lipid-breaking or lipolytic (*lipo*: lipid; *lytic*: breaking) enzymes collectively called 'lipases' (Figure 17). The triglycerides are sequentially degraded to diglycerides, monoglycerides and finally to free fatty acids and glycerol.

(*c*) Nucleolysis

During autolysis, the nucleic acids (DNA & RNA) of fish tissues are hydrolysed by a group of endogenous enzymes collectively called 'nucleases' (Figure 18). DNA is hydrolysed by the enzyme deoxyribonuclease (DNase) and RNA by the enzyme ribonuclease (RNase). They are hydrolysed to nucleotides.

4. Effects of Autolysis

Autolysis has both negative and positive effects. In fresh fish, autolysis has more pronounced negative effect on the quality. On the other hand, there are some processing methods, say preparation of fish silage and fish hydrolysate, where autolysis has more pronounced positive effect. However, the negative effects of autolysis are of greater concern than the positive effects.

(a) Negative Effects

(i) Softening of Tissue (Loss of Firmness)

The fish becomes soft and the firm elastic texture disappears. Now, the fish retains finger indentations.

(ii) Penetration of Microbes

After death, when the digestive enzymes of the gut hydrolyse the gut wall, the microbial flora penetrates through it into the hitherto sterile muscle and internal organs. This microbial flora acts on the muscle and internal organs and putrefies them (Figure 19). Similarly, when the components of the skin are hydrolysed, the microbial flora present on the skin penetrates through it into the muscle and internal organs and putrefies them.

(iii) Growth of Microbes

During autolysis, the complex tissue components are hydrolysed to simpler substances, which become a nutrient-rich medium for microbial growth. The microbes utilise these readily available simpler substances for their growth and metabolise them to off-odour, off-flavour substances presence of which renders the fish spoiled.

(iv) Increase in NPN Content

During autolysis, due to hydrolysis of proteins, substantial quantity of free amino acids is produced (Figure 15). This increases the non-protein nitrogen (NPN) content in fish tissue.

(v) Development of Rancidity

The hydrolysis of fish lipids results in the formation of free fatty acids (FFA) as one of the end products (Figure 17). Its accumulation in fish tissue results in the development of a type of rancidity called 'hydrolytic rancidity'.

(vi) Belly Bursting

It is a common post harvest problem in some species of pelagic fishes like oil sardine, herring, capelin and sprats. These species, during a particular period, usually in the post-spawning seasons, show high intake of food. If these fishes are caught during such periods of high food intake, the large quantities of enzymes secreted into the gut remain active for quite a long period of time after death. They degrade the gut wall completely within few hours and enter into the muscle. First, they hydrolyse the muscle adjacent to the gut. The end products of the hydrolysis of gut and adjacent muscle are simple substances like amino acids, which become a nutrient-rich medium for the activity of the natural microbial flora of the gut. Besides, the gut of fish having high intake of food contains a very high load of microbes, which have entered into the gut alongwith the food. All these microbes grow on the end products of hydrolysis resulting in the production of gases like CO_2 and H_2. The accumulation of these gases inside the belly cavity leads to the swelling of the belly. The belly, on the other hand, becomes soft due to the hydrolysis of its tissues by the gut enzymes. Ultimately the belly bursts after a short period from death. This phenomenon is called 'belly bursting'. The problem of belly bursting can be overcome by adopting certain measures. (1) By gutting the fish immediately after harvesting (2) By keeping the fish alive after harvesting so as to allow the ingested food to pass out from the gut as feces. (3) By quick chilling of fish immediately after harvesting. It is the most effective method of preventing belly bursting.

(vii) Gaping

The whole muscle of fish is supported by the collagenous connective tissue, myocommata. During autolysis, the proteolytic enzyme collagenase breaks down the protein collagen resulting in the disintegration of the myocommata. This results in separation of the myotomes into flakes. This phenomenon is called 'gaping'. Gaping results in its poor appearance resulting in consumers' dislikingness (p. 27).

A fish, which has passed through autolysis, is still acceptable to the consumer. However, the end products of autolysis become a nutrient-rich medium for the microbes. They utilise these simple end products and convert them to off-odour, off-flavour substances responsible for the spoilage of fish. Therefore, autolysis, though not spoilage in itself, leads to easy spoilage by microbes.

(*b*) Positive Effects

(i) Taste and Flavour Development

The intermediate products of nucleic acid and nucleotide break down during autolysis include AMP, IMP and GMP. The products of protein hydrolysis include amino acids like histidine, glutamic acid, glycine, valine, alanine, arginine and methionine. All these break down products impart pleasant flavour and taste to fish.

(ii) Liquefaction

The hydrolysis of fish tissues during the production of fish silage is enhanced by adding acids, which lower the pH of the fish tissue to the point favourable for the activity of the endogenous tissue enzymes.

(iii) Maturation of Salted Fish

In the temperate Scandinavian countries, salted fish is stored at cooler temperatures for months together (sometimes upto nine months) to undergo the process of 'maturation' or 'ripening'. During this period, the endogenous tissue enzymes hydrolyse the tissue components to simpler compounds, which impart the typical pleasant mature taste greatly relished by the people of these countries.

(iv) Tenderisation

Autolytic hydrolysis of the fish proteins results in the tenderisation of fish meat. Tenderised meat is soft and pliable for which it is organoleptically better accepted than the tough pre-tenderised meat.

5. Measurement of Autolysis

The extent of autolysis undergone after death is measured by estimating the concentration of the end products of autolysis such as free amino acids and free fatty acids. Their concentration is measured in terms of the following parameters.

(*a*) Alpha Amino Nitrogen (AAN)

It is an estimate of the concentration of free amino acids present in the fish tissue.

(*b*) Non-Protein Nitrogen (NPN)

When proteins are hydrolysed to amino acids, the free amino acids produced no longer remain part of the proteins and therefore count towards non-protein component. Fish tissues also contain other non-protein nitrogenous substances such as TMAO, creatine, urea etc. Thus the quantity of non-protein nitrogenous substances in fish expressed in terms of non-protein nitrogen (NPN) partly indicates the extent of autolytic hydrolysis of proteins.

(*c*) Free Fatty Acids (FFA)

During autolysis, fat is hydrolysed to free fatty acids and glycerol. The amount of free fatty acids (FFA) in fish tissues indicates the extent of hydrolysis of fats.

6. Control of Autolysis

The rate of autolytic degradation of tissue components is influenced by several factors such as temperature, presence of gut and internal organs, moisture content, salt content and pH of the fish tissue. If it is required to retard autolysis, then the steps to be taken include evisceration of the fish immediately after harvesting, exposure to low temperature by chilling or freezing, exposure to high temperature by cooking, canning or frying, subjecting the fish to drying and/or salting.

On the other hand, autolysis can be enhanced, if required, by lowering the pH through addition of acids to the point, which favours maximum activity of the autolytic tissue enzymes. Autolysis can also be accelerated by exposing fish to high temperatures not reaching the limiting temperature of enzyme denaturation.

5

Microbial Putrefaction

I. What is Microbial Putrefaction ?

After death of fish, rigor mortis sets in during which the body gets stiffened. Once rigor is resolved, the subsequent post mortem changes such as autolysis and microbial putrefaction start almost simultaneously. Due to autolysis, the complex tissue components are hydrolysed by the body enzymes to simpler substances. The firm elastic texture of fish body changes into soft flabby texture due to which it retains finger indentations. The end products of autolysis do not impart any remarkable off-flavour and off-odour to fish. A fish, which has passed through autolysis, is still acceptable to the consumer. Thus, autolysis alone will not spoil fish, if microbial activity can be arrested. However, autolysis helps in the process of spoilage by providing the microbes with a nutrient-rich medium consisting of the simple hydrolysed tissue components. Microbes also secret their own enzymes to the surrounding to hydrolyse the tissue components into simpler substances. The simple substances produced by autolysis as well those produced by the microbial hydrolysis of the tissue components are absorbed into the microbial cells and metabolised to the off-odour, off-flavour substances. It is the accumulation of these off-odour, off-flavour substances in the tissues, which renders the fish spoiled. 'Microbial putrefaction' or 'microbial decomposition' is the process of conversion of fish tissue components by microorganisms into off-odour, off-flavour substances accumulation of which distracts the consumers and thereby results in the spoilage of fish.

Microbes are the most important agent of fish spoilage. The next agent of importance is the enzymes. The third agent of importance is autoxidation, which has negligible role in the spoilage of fresh fish, but has the most important role in the quality loss of frozen fish. Although autolysis and microbial putrefaction set in almost simultaneously, depending on the storage temperature, one of them predominates initially. For example, when fish is preserved at a low temperature, say by icing, autolysis predominates initially whereas at a higher temperature of storage, initially microbial putrefaction predominates.

II. Importance of Microbes in Fish Processing

Microbes are of great concern in fish processing due to the following reasons.

1. Small Generation Time

Most of the microorganisms have extremely small generation time. One bacteria can give rise to another bacteria in few minutes. In enumeration of bacteria, each bacteria inoculated on a nutrient agar plate produces a colony of bacteria consisting of thousands of the same bacteria in only a few hours (say in 24 hours). Few bacteria, when inoculated into a nutrient broth, grow and multiply so rapidly that within few hours, the clear liquid broth becomes turbid, as thousands of bacteria remain suspended in it.

2. Resistance

There are few microbes, which, unlike many other organisms, can resist extremely unfavourable conditions by forming resistant spores.

3. Invisibility

Since they cannot be seen with naked eye, their presence is very often underestimated.

4. Agent of Fish Spoilage

They are the most important agent of fish spoilage.

5. Agent of Health Hazards

Some of the microbes are pathogenic whose presence in fish causes health hazards in persons who consume it.

III. Types of Microbes in Fish

Microorganisms or microbes are minute organisms mainly consisting of bacteria, fungi, algae, protozoa and virus. Among them, bacteria play the most important role in the spoilage of fish and in causing health hazards from consumption of fish. The microbes associated with fish spoilage and health hazards are of two types as follows.

A. Endogenous microbes

B. Exogenous microbes

A. Endogenous Microbes

Endogenous microbes, also called 'natural microbial flora' or 'resident microbes' are those microbes, which inhabit the fish body when fish is alive, but normally do not cause any disease in it. These microbes inhabit those parts of the fish body, which are exposed to the surrounding (Figure 20).

Fish body can be considered as a double-walled structure with the inner wall formed by the membranous lining of the gut and the outer wall consisting of the membranous lining of skin and gills. Both the membranous linings are impermeable to microbes when fish is alive. In between the two walls are found the muscle and internal organs. Thus, they are not exposed to the surrounding. Though gut remains within the fish body, it is exposed to the surrounding through its mouth at one end and cloacal aperture at the other. Microbes inhabit the skin, gills and gut in the increasing order of their load while muscle and internal organs are sterile (free of microbes) in healthy fish. In diseased fish microbes may invade the muscle and internal organs. The relative microbial load in these sites is given in Table 4 with examples.

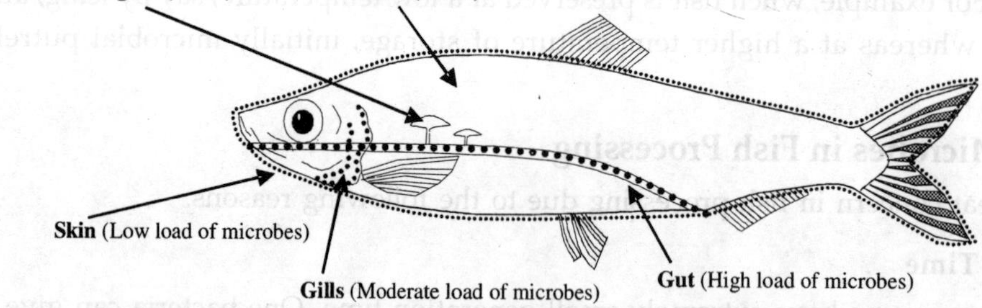

Figure 20: Natural microbial flora of fish

TEXT BOX 4

Microorganisms and life on the earth

Microorganisms or microbes are living forms whose size is so small that they are usually not visible to naked eye and can be seen only with the help of a microscope. They include bacteria, viruses, protozoa, microscopic fungi and algae. They are present in almost all places on the earth. Despite their minute size, they have tremendous importance in the maintenance of life on the earth. Due to the difference in the activities of different microbes, they influence life in different ways. Some of them are very helpful, for example, those which fix atmospheric nitrogen into biologically useful forms, those which help in curd formation, those which recycle dead materials by degrading them into simpler substances and those which help in preparation of wine. Some other microbes are very harmful in that they cause diseases in plants and animals, spoil food or raw materials of foods kept in normal conditions as well as they degrade rubber, paints, textiles, metals and even insulations on electric wires.

Based on their activity, microbes are, in general, categorised as follows.

A. Helpful Microbes

They help man in some way or the other. For example, the bacteria which help in the formation of curd from milk, the bacteria which help in production of alcohol by fermentation and the bacteria which help in the fixation of atmospheric nitrogen into organic nitrogen.

B. Harmful Microbes

They cause different types of harms to man. With respect to the use of fish as a food commodity, harmful microbes are of two types as follows.

(*a*) Spoilage Microbes (Putrefactive Microbes)

They produce some off-odour, off-flavour substances in fish, accumulation of which distracts the consumer and renders the fish spoiled. However, if consumed, it may not lead to any health hazard.

(*b*) Pathogenic Microbes

Pathogenic microbes are those whose presence or growth in fish can cause health hazard to the person who consumes it. The health hazard caused by the pathogenic microbes from consumption of fish comes under 'food poisoning'. There are two modes of food poisoning as follows.

1. Infection-type food poisoning (by endotoxins)
2. Intoxication-type food poisoning (by exotoxins)

1. Infection-type Food Poisoning

Food poisoning is said to be infection-type, if the food is infected with certain types of living food poisoning microorganisms. If these microorganisms are removed by some means or are killed by some method before consumption, food poisoning will not occur. In other words, when the presence of living microbes is essential for the poisoning, it is called infection-type food poisoning. These microbes possess 'endotoxin' (*endo*: within) in their cells. Endotoxin is nothing but the integral part of the cell wall of infectious microbes consisting of Lipid A fraction of the lipopolysaccharide (LPS) present in the outer membrane of gram-negative bacteria. Thus, a cell-free extract of the infected food will not cause poisoning as presence and activity of living microbes is essential for causing infection-type food poisoning. Microbes causing infection-type food poisoning include *Salmonella, Shigella, Escherichia coli, Vibrio cholerae, V. parahemolyticus, Aeromonas hydrophila, Listeria monocytogenes* etc.

2. Intoxication-type Food Poisoning

Food poisoning is said to be intoxication-type, if certain types of food poisoning microbes have grown in the food and have produced toxins in it. Even if these microorganisms are removed by some means or are killed by some method before consumption, still then food poisoning will occur. In other words, when the presence of living microbes is not essential for the poisoning as it is actually caused by the toxins produced by the microbes, it is called intoxication-type food poisoning. These microbes produce 'exotoxin' (*exo* : outside) which are released to outside from their cells. The exotoxins are proteinaceous substances synthesised within the cells of certain types of bacteria and released to outside. Thus, a cell-free extract of the infected food will cause poisoning as presence of toxin produced by microbes alone can cause intoxication type food poisoning. Microbes causing intoxication type food poisoning include *Clostridium botulinum, Staphylococcus aureus* etc.

Development of Endogenous Microbes in Fish

In a freshly hatched fish hatchling, the gut is sterile. As the hatchling grows, it takes in food from the surrounding. This food is loaded with microbes. In this way, microbes of the surrounding enter into

Table 4: Natural microbial flora in different parts of fish body

Site in Fish Body	*Relative Microbial Load*	*Examples*		
		Rohu (Freshwater Fish)	*Oil Sardine (Marine Fish)*	*Pearl Spot (Brackishwater Fish)*
Gut (intestine)	High	10^5–10^7	10^5–10^9	10^5–10^8
Gills	Medium	10^4–10^6	10^5–10^8	10^4–10^8
Skin	Low	10^4–10^5	10^3–10^7	10^3–10^4
Muscle and internal organs	Nil (Sterile)	Nil	Nil	Nil

the gut alongwith the food. As the food passes through the gut, it is hydrolysed by the gut enzymes to simpler forms. A large portion of it is absorbed through the gut wall into the blood stream and a small residual portion is utilised by the microbes, which enter the gut alongwith the food, for their growth and multiplication. Besides, the microbes themselves hydrolyse the undigested food by their own enzymes and utilise them. In this way, the microbes get access into the gut and colonise there as the 'natural microbial flora of the gut'[1]. The microbial flora of the gills develops in a similar manner. Fish passes huge quantities of surrounding water over its gills for its respiration. The microbes present in this water get filtered on the gills and inhabit there as the 'natural microbial flora of the gills'. Similarly, the microbes present in the surrounding water adhere to the skin and become the 'natural microbial flora of the skin'. The natural microbial flora of fish is of two types as follows.

1. Spoilage microbes
2. Pathogenic microbes

Type and Density of Endogenous Microbes

The natural microbial flora of fish caught from open waters (*e.g.* open seas), which are away from human activities, mostly consists of 'spoilage microbes'. These microbes act on tissue components of fish after death and metabolise them to off-odour, off-flavour substances presence of which renders the fish spoiled. The pathogenic microbes inhabit the fish body only when it comes in contact with polluted waters or waters with intense human activities such as ponds, tanks, lakes etc. They cause health hazards when fish contaminated with these microbes is consumed. The type and density of microbes that constitutes the natural microbial flora depends mainly of the following factors.

1. *The Habitat of Fish*

The natural microbial flora of fishes of marine habitats mostly consists of gram-negative, highly salt tolerant, spoilage bacteria (Table 5). Pathogenic microbes are mostly absent as most of the marine habitats such as the open seas are unpolluted and are far from human activities. The natural microbial flora of fishes of freshwater habitats are mostly a mixture of gram-positive and gram-negative bacteria which are non-salt tolerant and include both spoilage and pathogenic microbes. The pathogenic microbes are present in the fish body as most of the freshwater habitats (ponds, lakes, ditches, rivers etc.) are exposed to contamination from intense human activities and pollution. The natural microbial flora of fishes of brackishwater habitats are mostly a mixture of gram-positive and gram-negative bacteria which are moderately salt tolerant and contain spoilage microbes with rare pathogens. This is because; most of the human pathogens cannot tolerate even the moderate salinity of brackishwater.

2. *The Extent of Pollution and Exposure of the Habitat to Human Activities*

When fish inhabits unpolluted waters or waters which remain away from human activities (*e.g.* the open seas), its natural microbial flora consists mostly of spoilage microbes. The pathogenic microbes

1 The natural microbial flora of the gut is saprophytic in that, it lives on dead materials such as the undigested food that passes down the gut. In higher animals, this flora is helpful as it produce some of the vitamins essential for the body. When antibiotics are used against any disease, it may kill the harmless gut microbes too besides killing the targeted pathogens. This may lead to vitamin deficiency diseases. Therefore, usually vitamins are taken alongwith antibiotics during treatment of diseases.

Table 5: Natural microbial flora of fishes of different habitats

Marine Habitat	*Freshwater Habitat*	*Brackishwater Habitat*
Mostly gram-negative	A mixture of gram-negative and gram-positive bacteria	A mixture of gram negative and gram-positive bacteria
Highly salt tolerant	Non-salt tolerant	Moderately salt tolerant
Mostly spoilage type microbes	A mixture of both spoilage and pathogenic microbes	Mostly spoilage type microbes with rare pathogens.
Polluted and exposed to human activities : Pathogens + spoilage microbes Unpolluted and not exposed to human activities : No pathogens + only spoilage microbes		

get their access to the fish body only when fish inhabits polluted waters (*e.g.* industrial sewage) and waters exposed to intense human activities (the inland water bodies like lakes, ponds, ditches and rivers) which are rich in pathogenic microbes. Fishes of coastal waters also harbour high load of pathogenic microbes due to land drainage into the seas.

3. *Location of Waterbody from which Fish is Caught*

The type of microbes found in fish also depends on the location of waterbody from which it is caught. A fish caught in the North Sea has a characteristic microbial flora different from that caught in the Mediterranean Sea and a fish caught in the Arabian Sea has a microbial flora different from that caught in the Bay of Bengal.

B. Exogenous Microbes

They are the microbes, which get their access to the fish body during post harvest handling, processing and storage. Their main sources are the twines of the nets, mud adhering to fish body and nets, the decks, the pound boards of fish-holds used for storing fish in fishing vessels, cutting boards, washing tubs, water used for washing fish, ice used for chilling fish, floor of the processing halls, processing tables, knives, utensils and above all, the hands of the processors. Similar to endogenous microbes, the exogenous microbes are also of the same two types as follows.

1. Spoilage microbes
2. Pathogenic microbes

IV. Species of Microbes Associated with Spoilage of Fish

It is important to note that generally the endogenous microbes or the natural microbial flora present in fish are spoilage type while the exogenous microbes contaminating the fish during post harvest handling, processing and storage are pathogenic type. The different species of microbes associated with the spoilage of fish through microbial putrefaction are given in Table 6. The pathogenic microbes have no remarkable role in the post mortem microbial putrefaction of fish and are therefore not discussed here.

Table 6: Microbes associated with spoilage of fresh fish

Pseudomonas spp.	*Micrococcus* spp.
Aeromonas spp.	*Corynebacterium* spp.
Achromobacter spp.	*Photobacterium* spp.
Flavobacterium spp.	*Bacillus* spp.
Vibrio spp.	*Shewanella* (formerly *Altromonas*) spp.
Moraxella spp.	*Proteus* spp.
Arthrobacter spp.	

V. Spread of Microbes in the Body of Dead Fish

The microbes are unable to spread into the sterile muscle and internal organs of live fish due to (1) Presence of defense mechanism or immunity of body, (2) Presence of microbe-impermeable membranous linings (gut as the inner lining and skin as the outer lining) and (3) Lack of readily utilisable nutrients for microbial growth and multiplication. However, after death of fish, they penetrate into the muscle and internal organs due to the following reasons and putrefy them.

1. Loss of defense mechanism or immunity of body.
2. Loss of microbe-impermeability of the membranous linings of the body.
3. Abundance of readily utilisable nutrients for microbial growth and multiplication.

1. Loss of Defense Mechanism or Immunity of Body

The body of living fish has an effective defense mechanism against invasion by microorganisms. This defense mechanism is called 'immunity'. Due to the existence of immunity in fish as well as in all other organisms, they are not succumbed to frequent diseases despite their constant exposure to pathogenic microbes, which are present in all the materials they come in contact throughout their life. This immunity is offered by two agents. (1) The white blood cells (WBC or leucocytes) floating in the blood engulf the infectious cells and destroy them. The WBCs are circulated in the body via the blood circulatory system. (2) The antibodies are special protein particles called 'immunoglobins' produced by the organism specifically against invasion by foreign particles such as viruses. The foreign particles are called antigens. A specific antibody is produced against a specific antigen. Thus, an antibody produced against a particular antigen cannot neutralise any other antigen. The antibody binds to the antigen producing an 'antigen-antibody complex'. Once entrapped in the complex, the antigen cannot cause the disease. In this way, the antibody neutralises the antigen. The huge surplus of antibodies produced against a particular antigen is circulated throughout the body via blood circulatory system, which protects the fish body against invasion, by the same antigen in subsequent times.

After death of fish, the immunity of fish body is lost due to which microbes invade the fish tissues, proliferate there and spoil it. Immunity is lost due to the following reasons.

(*a*) Production of fresh antibodies (immunoglobins) stops after death.

(*b*) Oxygen supply to blood stops after death resulting in the death of the WBCs.

(*c*) Due to termination of blood circulation after death, the WBC and the antibodies already present in the blood cannot reach different parts of the body.

2. Loss of Microbe-Impermeability of the Membranous Linings of the Body

Fish body can be considered as a double-walled structure with the inner wall formed by the membranous lining of the gut and the outer wall consisting of the membranous lining of skin and gills. Both the membranous linings are impermeable to microbes when fish is alive. In between the two walls are found the muscle and internal organs. Thus, they are not exposed to the surrounding. Though gut remains within the fish body, it is exposed to the surrounding through its mouth at one end and cloacal aperture at the other. Microbes inhabit the skin, gills and gut in the increasing order of their load while muscle and internal organs are sterile (free of microbes) in healthy fish.

However, after death of fish, due to autolytic hydrolysis the microbe-impermeability of the membranous linings is damaged thereby allowing the microbes to penetrate into the sterile muscle and internal organs and putrefy them. There are three routes through which microbes penetrate into the muscle and internal organs as describes below.

(*a*) Gut

After the death of fish, autolysis results in the break down of fish tissue components, particularly the proteins, by its own body enzymes. The gut enzymes, which remain active long after death of fish, hydrolyse the gut wall and it becomes porous for the microbes to penetrate. The gut microbes penetrate through the porous gut wall into the hitherto sterile muscle and internal organs, grow there and putrefy them.

(*b*) Gills

The gills are soft, moist and rich in blood. It is bright red in colour and has no offensive smell. After death, the natural microbial flora of the gills enters into the soft gill tissue and grows there. As the microbial load of the gills is high, these microbes quickly produce remarkable changes in the gills manifested by change of colour from bright red in fresh fish to dark brown in spoiled fish associated with production of offensive smell. A consumer assumes a fish to be fresh if the colour of its gills is bright red and it has no off-odour. In this way, the consumer often uses the state of the gills as an indicator of freshness of fish.

The natural microbial flora of gills also moves rapidly from the gills to the muscle and internal organs via the blood circulatory system, which passes through the gills. In order to prevent this, it is important to bleed larger fishes as soon as possible after harvesting.

(*c*) Skin

When fish is alive, a thin layer of slime covers its body. It protects the fish body against microbial invasion in three ways as below.

1. In some fishes the slime contains antibodies, which have immunological properties against microbial invasion into the fish body.
2. In some fishes the slime contains the antimicrobial substance 'lysozyme' that can kill bacteria.
3. The slime layer acts as a physical barrier against microbial attack.

But when fish dies, the following events occur.

1. The immunological properties of the slime are lost and production of fresh antibodies stops.
2. Microbes attach to the sticky slime and do not get washed away easily.
3. Slime consists mainly of the glycoprotein, mucin, which is a rich nutrient for microbes. It becomes an excellent substrate for microbial growth. The natural microbial flora of the skin grows in the slime rapidly after death.

Due to the above events, the microbes grow profusely in the slime and putrefy it. As a result, it gives off an offensive smell. Besides, the profuse growth of the skin microbes increases the chance of their penetration into the sterile muscle and internal organs. They penetrate through the skin when it becomes permeable to them upon its hydrolysis by the body enzymes during autolysis. Their entry becomes still easier, if it gets cuts or bruises during handling through which they can enter into the muscle and internal organs. They grow there profusely and putrefy them.

3. Abundance of Readily Utilisable Nutrients for Microbial Growth and Multiplication

It is mainly the nitrogenous substances present in fish tissues whose decomposition, after death, by the microbes, results in the production of off-odour, off-flavour substances leading to the spoilage of fish. The important nitrogenous constituents of the fish tissues are proteins, free amino acids, TMAO, urea, creatine, creatinine and nucleic acids. Except proteins, which contribute to the bulk of the nitrogenous substances, all other nitrogenous substances are called non-protein nitrogenous (NPN) substances. Proteins are large macromolecules due to which they cannot enter into the cells of microbes to be utilised by them. Therefore, to utilise the proteins, microbes have to secret proteolytic enzymes to the surrounding where they hydrolyse the proteins into amino acids. These free amino acids are very small molecules and are absorbed into the microbial cell where they are metabolised to off-odour, off-flavour substances. Such utilisation of macromolecular nitrogenous substances like proteins is a slow process. Therefore, NPN, which can be directly absorbed into the microbial cells due to their smaller size, are the best substrates for the growth of microbes.

The microbes at first utilise the NPN such as TMAO, urea, nucleic acids and the free amino acids naturally available in fish tissues when it is alive. In the tissue of living fish, the concentration of NPN is very low. However, during autolysis, the tissue proteins are hydrolysed by the body enzymes to free amino acids, which become an abundant readily utilisable nutrient for microbial growth and multiplication. The microbes absorb and metabolise them easily in short time to off-odour, off-flavour substances, the presence of which renders the fish spoiled. Once these NPN are exhausted, the microbes begin to secret proteolytic enzymes, which attack tissue proteins and hydrolyse them to free amino acids for their subsequent absorption and utilisation by these microbes.

It is important to note that, the rate of autolysis is practically same for all proteins. Therefore, the concentration of amino acids produced by autolytic hydrolysis of proteins is almost same for all proteins. Thus, it is the concentration of NPN present in live fish, which determines the rate of microbial decomposition. Marine fishes have higher NPN due to their higher content of TMAO (in teleosts) and urea (in elasmobranchs). Therefore, they take less time for microbial decomposition and spoilage as compared to freshwater fishes, which have less NPN. That is why; marine fishes usually spoil faster than freshwater fishes. The spoilage of fish with large content of histidine, which is an amino acid, is particularly rapid.

VI. Mechanism of Microbial Putrefaction

Microbes decompose and spoil fish in four ways as given below.

1. Utilisation of readily utilisable simple substances such as NPN (non-protein nitrogenous) substances
2. Hydrolysis of complex tissue components (*e.g.* proteins) into simpler substances (*e.g.* amino acids) and their subsequent utilisation.
3. Rancification
4. Nucleic acid decomposition

1. Utilisation of Readily Utilisable Simple Substances such as NPN

The first substances to be utilised by the microbes are the readily utilisable simple substances, mainly the NPN components. They include

A. TMAO (in marine fishes and shellfishes)

B. Urea (in elasmobranchs)

C. Free amino acids (in all fishes)

A. TMAO Degradation

TMAO (Trimethylamine oxide) is an osmo-regulatory substance present in the tissue of marine fishes and shellfishes. Freshwater fishes generally contain negligible quantities of TMAO. Among marine fishes, elasmobranchs contain the highest and flat fishes the lowest amount of TMAO. Pelagic fishes and cod family contain TMAO in moderate quantities. The high content of TMAO in shark (no less than 2.5 per cent on dry weight basis) is due to its role in the osmo-regulation of this fish. In muscle of shark, about 750-1480 mg per cent TMAO is found. Urea and TMAO comprise 60-76 per cent of the nitrogen extractives in shark muscle. White-fleshed demersal fishes generally contain large quantities of TMAO than dark-fleshed pelagic fishes. It is sequentially degraded through reduction to trimethylamine (TMA), dimethylamine (DMA), methylamine (MA) and finally to ammonia (NH_3). Reduction of TMA involves reduction of one methyl group in each step with its liberation as methane (CH_4) (Figure 21).

Among all the steps, the first step in which TMAO is reduced (by addition of H_2) to TMA is most important. TMAO is mostly reduced to TMA enzymatically by the enzyme 'TMAO reductase' in which cytochrome acts as electron carrier. Enzymatic reduction, though mostly performed by microbial enzyme, also takes place by tissue enzymes of fish. Non-enzymatic reduction, which is less important, takes place in presence of certain catalysts or by heating of various seafoods during processing (Figure 22).

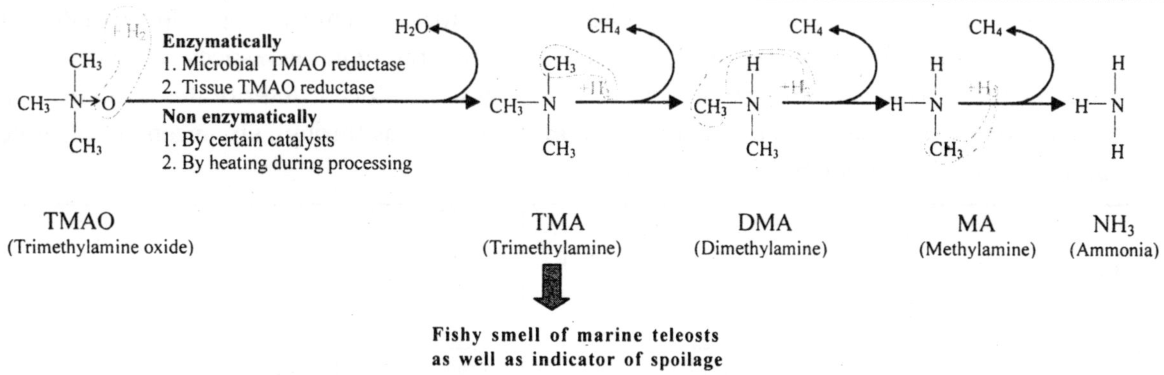

Figure 21: Degradation of TMAO in fish tissues after death

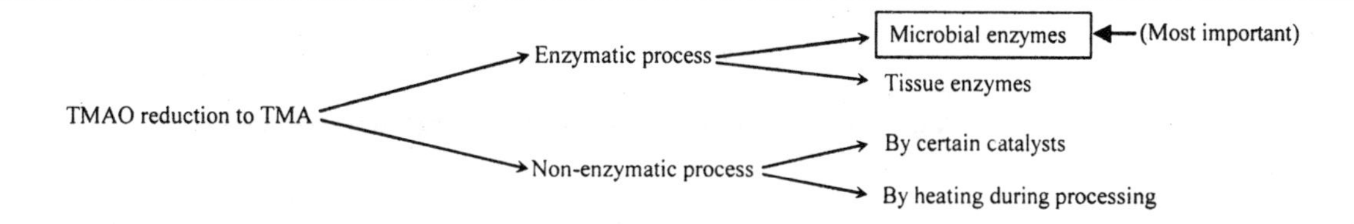

Figure 22: Different routes of TMAO degradation

Among the break down products of TMAO, TMA is most important. It is a strong smelly substance and is responsible for the typical fishy odour of marine teleosts. The concentration of TMA in the tissue of marine fishes is estimated by determining its nitrogen (N) content and is expressed as TMAN (trimethylamine nitrogen) *i.e.* nitrogen in TMA. As TMAO is degraded to TMA mostly by microbial action, the content of TMA in fish tissue, expressed as TMAN, is used as an indicator of the extent of microbial spoilage of marine fishes and shellfishes. TMAN reflects how much of TMA has been formed in fish tissue from TMAO by its microbial degradation. Thus, the content of TMAN in tissue increases with increase in microbial spoilage. However, as TMA is an intermediate product and is degraded further, its concentration at any instance may not indicate exactly how much of TMAO has been degraded by the microbes. Moreover, TMAN cannot be used as an indicator of spoilage of freshwater fishes, which lack TMAO.

The break down products of TMAO are the simple monoamines (*i.e.* containing one amine group, –N=) such as trimethylamine (TMA), dimethylamine (DMA), methylamine (MA) and ammonia (NH_3). In freshwater fishes, which lack TMAO, ammonia is produced by the deamination of the amino acids by microbes (Figure 26). All these compounds are volatile, which escape from fish body into the surrounding air (volatiles) and produce off-odour of spoiling fish. As they are basic (alkaline) in nature, they are collectively known as volatile base nitrogenous substances (VBN). The concentration of VBN in fish tissue is estimated by determining its nitrogen (N) content and is expressed as TVBN (total volatile base nitrogen) *i.e.* nitrogen in total volatile base nitrogenous substances (Figure 23). As VBN is produced mostly by microbial action, the content of VBN in fish tissue, expressed as TVBN, increases with increase in microbial action. Therefore, TVBN is used as an indicator of the extent of microbial spoilage of both marine as well as freshwater fishes. As it includes both, the end products as well as the intermediate products, in case of marine fishes its concentration, unlike TMAN, at any instant can indicate how much of TMAO has been degraded by microbial action. Therefore, TVBN is a better indicator of microbial spoilage of marine fishes than TMAN.

-TMA (Trimethylamine)
-DMA (Dimethylamine)
-MA (Methylamine)
-A (Ammonia)

} **Volatile bases** (Expressed as total volatile base nitrogen, TVBN)

Figure 23: The simple monoamines produced by microbial decomposition of TMAO

In the gadoid fishes like Alaska pollack, cod and haddock, TMAO is degraded to DMA and formaldehyde by 'TMAO demethylase', an endogenous enzyme present in these species. It occurs during cold storage of these fishes after freezing. The formaldehyde produced deteriorates protein quality during frozen storage of these fishes by denaturation of their proteins through enhancement of protein-protein and protein-lipid interaction. (Figure 24).

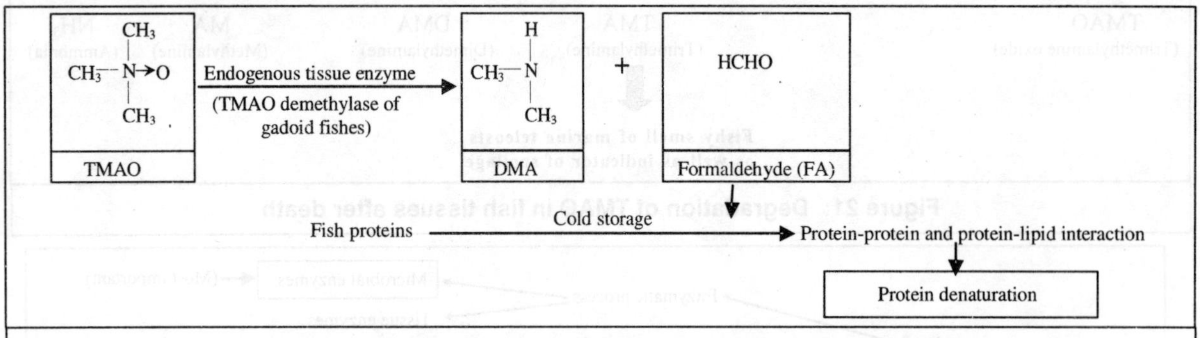

Figure 24: TMAO degradation in gadoid fish leading to protein denaturation

B. Urea Degradation

Urea is the osmo-regulatory substance present in the tissues of cartilaginous fishes like sharks, skates and rays. It is decomposed to carbon dioxide (CO_2) and ammonia (NH_3) by the microbial enzyme 'urease'. The ammonia imparts the ammoniacal odour of spoiling cartilaginous fishes (Figure 25).

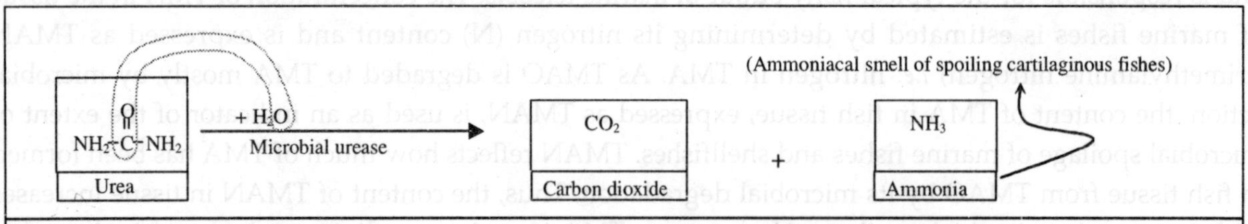

Figure 25: Microbial decomposition of urea in cartilaginous fishes

C. Free Amino Acid Degradation

The tissue of live fish contains non-protein nitrogenous substances (NPN) such as free amino acids, TMAO, urea etc. The free amino acids are present in tissue of live fish in very small quantity. However, after death, large quantity of free amino acids is produced by autolytic hydrolysis of tissue proteins resulting in an increase in the NPN value. It is these free amino acids, which make the fish body a nutrient-rich medium for microbial activity. Free amino acids are also produced from hydrolysis of tissue proteins by the proteolytic enzymes secreted by the microbes to the surrounding, when the available free amino acids is scanty or has been exhausted. The free amino acids are absorbed into the microbial cells where they are metabolised mostly through decarboxylation and deamination under specific enzyme systems of the microbes. The metabolic end products come out of the microbial cells. They have intense off-odour and off-flavour. Their presence in the fish tissue in high concentration renders the fish spoiled.

Degradation of the free amino acids takes place in three ways as given below.

(a) Deamination

It is the removal of the amino ($-NH_2$) group of the amino acids. The amino group so removed is reduced to ammonia (NH_3). Deamination occurs in four ways as follows (Figure 26).

(*i*) *Oxidative Deamination*: It takes place in presence of oxygen with production of a keto acid and ammonia. The keto acid eventually turns into aldehydes, lower fatty acids and CO_2 (Figure 26A).

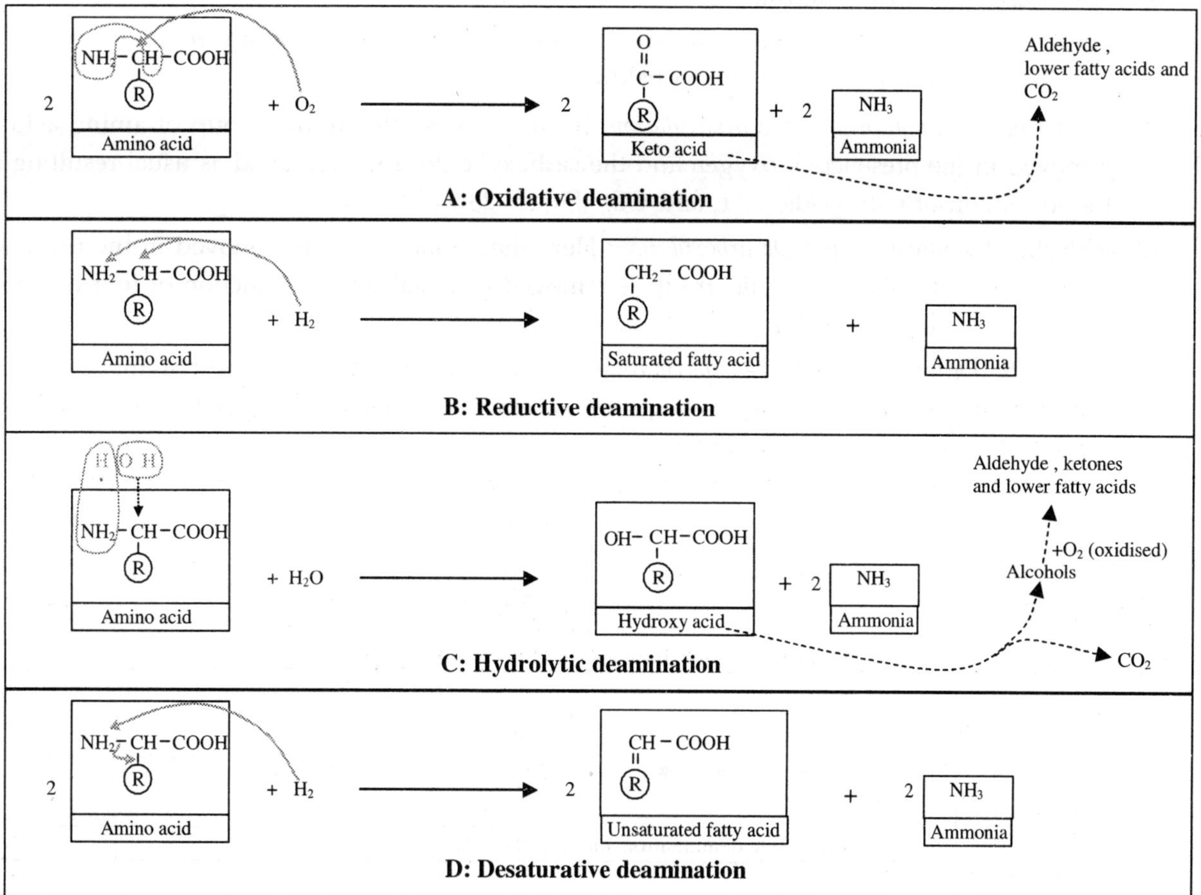

Figure 26: Deamination of amino acids by microbes in different ways

(*ii*) *Reductive Deamination*: It occurs in presence of hydrogen producing a saturated fatty acid and ammonia (Figure 26 B).

(*iii*) *Hydrolytic Deamination*: In this process free amino acids are hydrolysed in presence of water with production of a hydroxy acid and ammonia. The hydroxy acid (= oxy acid) is decarboxylated with liberation of CO_2 (from the –COOH group of the amino acids) and production of alcohols. The alcohol is oxidised further to aldehydes, ketones and lower fatty acids (Figure 26 C).

(*iv*) *Desaturative Deamination*: Here, the amino acid is converted to an unsaturated fatty acid with formation of ammonia (Figure 26 D).

(*b*) *Decarboxylation*

It is the removal of the carboxylic group (–COOH) from the amino acids with production of CO_2 and amines. These amines are called 'biogenic amines' and include histamine, cadaverine, putrescine, agmatine etc. (Figure 27).

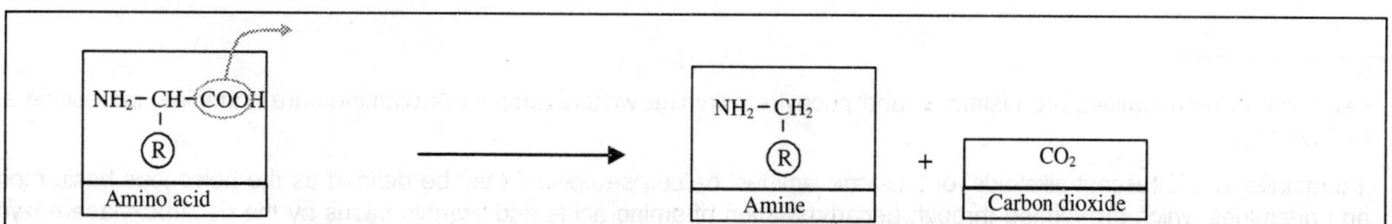

Figure 27: Decarboxylation of amino acids to produce biogenic amines

(c) *Deamination and Decarboxylation*

Here both, the amino group ($-NH_2$) and the carboxylic group (–COOH) are removed from the amino acids. It occurs in three ways as given below:

(i) *Oxidative Deamination and Decarboxylation*: In this process the amino group of amino acid is removed in the presence of oxygen and the carboxylic group is removed as usual resulting in the formation of fatty acids, NH_3 and CO_2 (Figure 28 A).

(ii) *Reductive Deamination and Decarboxylation*: Here the amino group is removed in the presence of hydrogen and the carboxylic group is removed as usual with production of hydrocarbon, NH_3 and CO_2 (Figure 28 B).

(iii) *Hydrolytic Deamination and Decarboxylation*: In this case, the amino group is removed in the presence of water and the carboxylic group is removed as usual with production of primary alcohol, NH_3 and CO_2 (Figure 28 C).

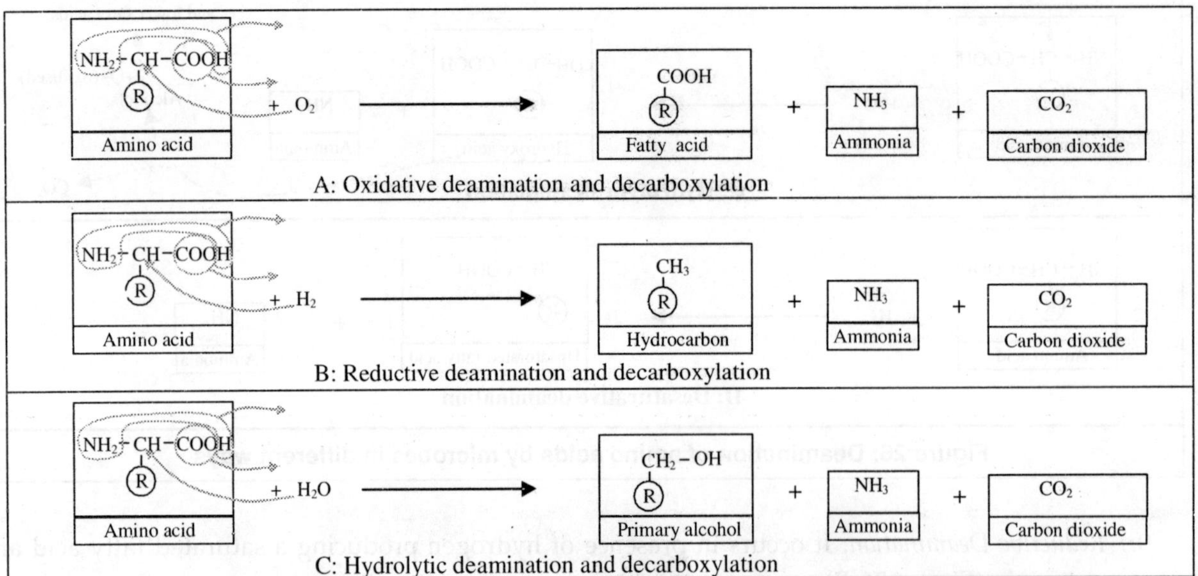

Figure 28: Deamination and decarboxylation of amino acids by microbes

The decomposition of important amino acids by microbes is described below individually.

I. Histidine

The amino acid histidine is a cyclic mono-amino acid; cyclic as it contains a ring structure and mono-amino as it contains one amino ($-HN_2$) group. It is present in extremely high levels in the muscle tissue of dark-fleshed fishes belonging to the family 'scombridae'. They include tuna, mackerel, seer fish etc. It is decarboxylated by microbial 'histidine decarboxylase' to produce the cyclic monoamine[2], 'histamine' and CO_2 (cyclic: containing ring structure, monoamine: containing one amino group). Histamine comes under the group 'ptomaines'[3] or 'putrescent alkaloids' or 'biogenic amines' as it is produced by bacterial putrefaction of amino acids through decarboxylation (Figure 29). Presence of histamine in fish and fishery products causes a type of poisoning called 'histamine poisoning' or 'scombroid poisoning' (Text box 5).

---

2 Examples of monoamines are histamine and phenyl ethylamine while examples of diamines are agmatine, putrescine and cadaverine.

3 'Ptomaines' or 'putrescent alkaloids' or 'biogenic amines' or 'corpse poison' may be defined as the poisonous basic mono- and di-amines which are formed through decarboxylation of amino acids and organic bases by the decarboxylase enzyme of putrefactive bacteria. The word 'ptomaines' was coined by the Italian toxicologist named Selmi in 1870 from the word 'ptoma', which means corpse or dead body.

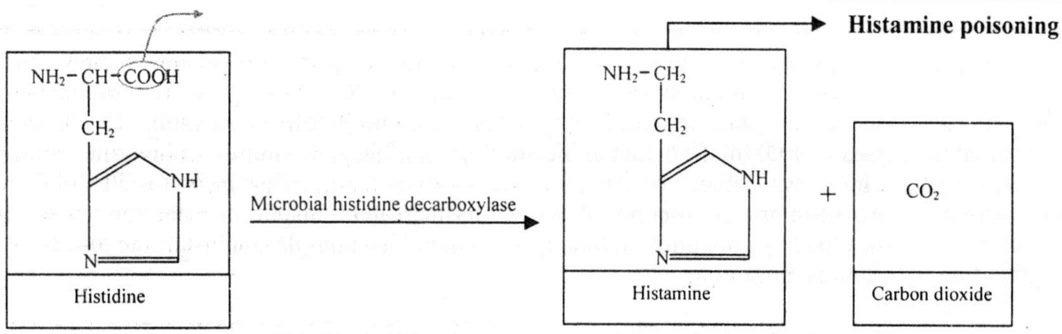

Figure 29: Microbial decarboxylation of histidine

TEXT BOX 5
Histamine poisoning

The presence of histamine in foods in high concentration causes a type of food poisoning called 'histamine poisoning' or 'histamine toxicity' when such food is consumed. It is also called 'scombroid poisoning' as it is mostly caused by the fishes belonging to the family 'scombridae', which have high histidine content in their tissues. These scombroid fishes include tuna, mackerel, bonito, seer fish etc. Other non-scombroid pelagic fishes such as sardine, anchovy, pilchard, jack mackerel, marlin and mahi mahi (dolphin fish) also cause histamine poisoning. On the contrary, fishes like herring produce histamine, but do not cause histamine poisoning. This is because, some other amines present only in scombroid fishes act as synergist for the poisoning action of histamine. These synergistic amines include putrescine and cadaverine, which are produced alongwith histamine by the decarboxylation of the amino acids ornithine and lysine respectively. In their presence, a lower level of histamine can cause poisoning as compared to the level of histamine required to cause poisoning alone. Several microbes are associated with the production of histamine from histidine in fish (Table 7).

Table 7: Microbes associated with histamine production in fish

Morganella morganii (previously *Proteus morganii*) (most important)	*Enterobacter cloacae*
Klebsiella pneumoniae	*Escherichia coli*
Hafnia alvei	*Edwardsiella* spp.
Citrobacter freundii	*Vibrio alginolyticus*
Clostridium perfringens	*Lactobacillus* spp.
Enterobacter aerogenes	

Histamine is an allergic substance (allergen), which produces allergic reactions in many people. Therefore, in histamine poisoning, an allergic-type condition occurs. It is a mild disease; incubation period is very short (5 minutes to 5 hours) and duration of illness is short (few hours). The most common symptoms are cutaneous (facial flushing, urticaria, edema), gastrointestinal (nausea, vomiting, severe stomach pain, diarrhoea) and neurological (headache, tingling, burning sensation in the mouth). Considering the severity of the poisoning, regulatory limits of histamine in fish and fishery products have been prescribed in many countries (Table 8).

Table 8: Regulatory limits of histamine in fish and fishery products in different countries

Country	*Limit (mg%)*	*Type of Limit*
USA (FDA i.e. Food and Drug Administration)	50	Hazard action level
	10-20	Defect action level
EEC (European Economic Community)	10	Defect action level
	20	Maximum allowable limit
Canada	>10	Indicates decomposition
	10-20	Defect action level
Germany	20	–
Denmark	30	–
India	20	–
Sweden	20	–

Although in all outbreaks of histamine poisoning, the food samples were found to contain high levels of histamine, it is important to note that presence of histamine in food does not always result in the disease even when hazard action level is exceeded. This is because; the human body can tolerate a certain amount of histamine

Contd...

Text Box 5–Contd...

without any reaction. The ingested histamine is detoxified in the intestinal tract by two enzymes namely 'diamines oxidase' (DAO) and 'histamine N-methyltransferase' (HMT) (Figure 30). These protective mechanisms get eliminated if; **(1)** These two enzymes are blocked by other compounds (drugs, alcohol), **(2)** There is any gastrointestinal tract disease or **(3)** Intake of histamine and/or other biogenic amines (cadaverine, putrescine, agmatine) is very high. In this way, other biogenic amines act as potentiators of histamine toxicity. Inhibition of intestinal detoxification of histamine as described above results in greater transport of histamine across cellular membranes and into the blood plasma through blood circulation. Elevated plasma histamine results in toxic reactions leading to histamine poisoning.

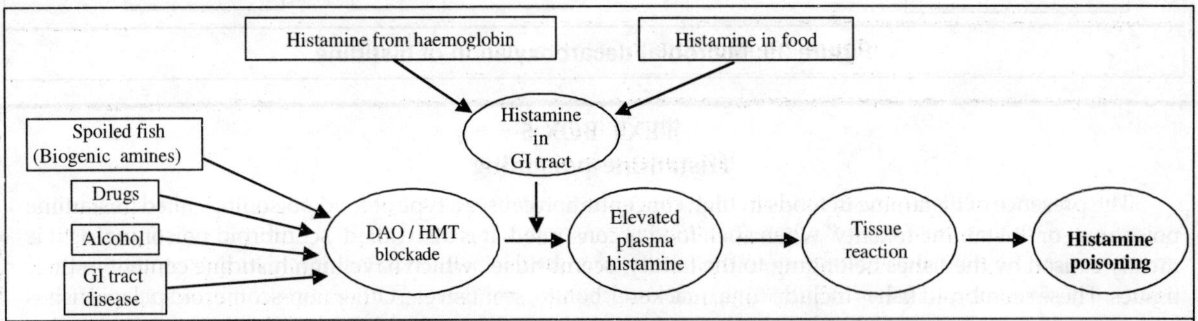

Figure 30: Pathway of histamine poisoning

Histamine, once produced in fish, can neither be removed nor damaged. It is heat stable and is retained even after heat processing during canning. Though histidine is degraded in different pathways (Text box 6), post mortem degradation of histidine to histamine in fish takes place only through bacterial decarboxylation. Thus, the histamine formation can be reduced by retarding the microbial activity. Therefore, histamine production in the fish tissue can be reduced by taking those steps, which are usually adopted to retard microbial activity. They include the following steps.

1. Quick chilling of fish immediately after harvesting by icing, chilled seawater stowage (CSW) or refrigerated seawater stowage (RSW).
2. Immediate freezing of fish.
3. Maintenance of good hygienic practices.
4. Processing fish as quickly as possible without allowing it to undergo microbial putrefaction.

II. Phenylalanine

The amino acid phenylalanine is a cyclic mono-amino acid; cyclic as it contains a ring structure and mono-amino as it contains one amino ($-HN_2$) group. It undergoes decarboxylation by microbial 'phenylalanine decarboxylase' to produce phenylethylamine and carbon dioxide (Figure 34). Like histamine, phenylethylamine is also a cyclic monoamine and comes under the group ptomaines or biogenic amines.

III. Lysine

The amino acid lysine is a non-cyclic diamino acid; non-cyclic, as it has no ring structure and diamino as it has two amino ($-NH_2$) groups. It is decarboxylated by microbial 'lysine decarboxylase' to produce cadaverine and carbon dioxide (Figure 35). Cadaverine is a non-cyclic diamine (*i.e.* no ring, but two amino groups) and is a ptomaine or biogenic amine.

IV. Arginine

It is decarboxylated by microbial 'arginine decarboxylase' to produce agmatine and carbon dioxide (Figure 36). Agmatine is a non-cyclic diamine (*i.e.* no ring, but two amino groups) and is a ptomaine or biogenic amine. Such a process is undergone if the microbe is *E. coli*. In another way, arginine is degraded by *Streptococcus faecalis* to urea and ornithine. Ornithine, in turn, is degraded by *E.coli* to putrescine.

TEXT BOX 6
Non-microbial degradation of histidine

Although histidine is degraded to histamine by microbial decarboxylase, the major pathway of histidine break down is deamination by the tissue enzyme 'histidine deaminase' producing glutamic acid and imino methylamine. This enzyme is present in the liver and kidney of most fishes. In mackerel, it is also present in the muscle. The enzyme actually does not remove the alpha amino group (*i.e.* the – NH_2 attached to the C to which –COOH is attached) of histidine; rather it reduces the –NH group in the ring structure to $-NH_2$, thereby liberating an amine alongwith glutamic acid (Figure 31). Another tissue deaminase called 'urocanase' deaminates histidine, by removal of the alpha amino group, thereby producing urocanic acid and ammonia (Figure 32). Degradation of histidine by the tissue enzyme histidine hydrolase produces formiminoglutamic acid and ammonia (Figure 33).

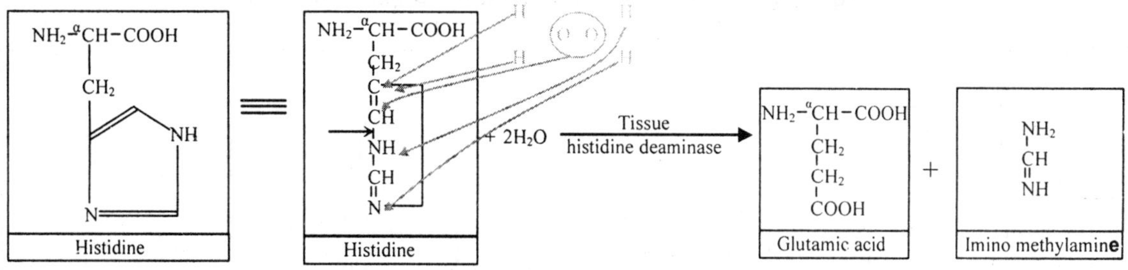

Figure 31: Histidine degradation by tissue histidine deaminase

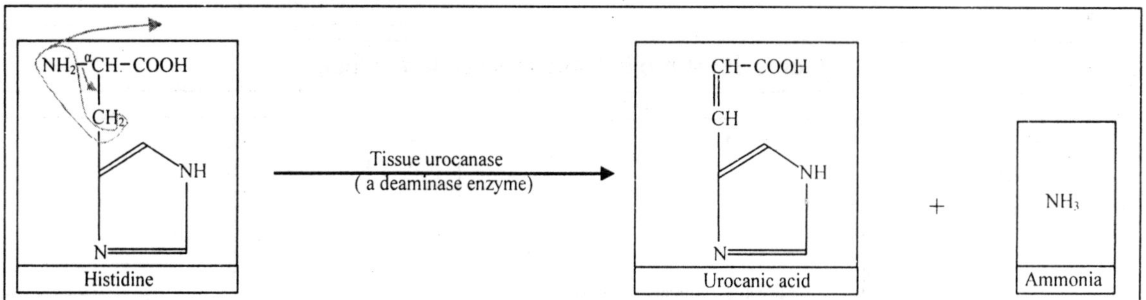

Figure 32: Histidine degradation by tissue urocanase

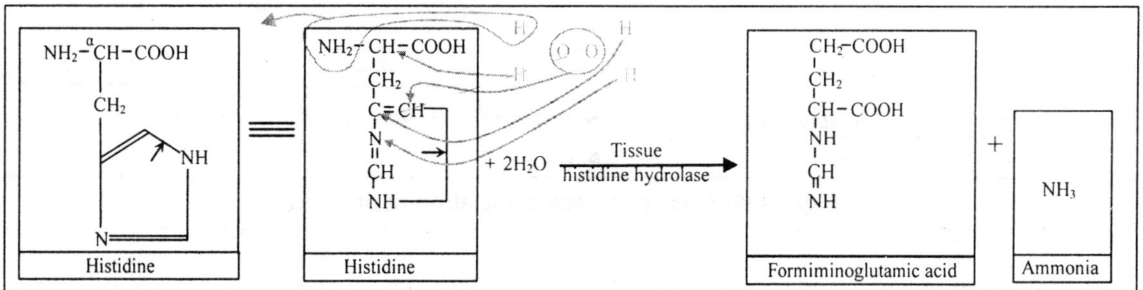

Figure 33: Histidine degradation by tissue histidine hydrolase

V. Ornithine

There are three sources of ornithine in fish tissue (Figure 37). (1) It is a non-protein amino acid as it does not form peptide linkage and is therefore not found in proteins. It is present in fish tissue as free ornithine. (2) Some microbes like *Streptococcus faecalis* degrade the amino acid arginine to ornithine with liberation of urea. (3) It is also produced by microbial degradation of the amino acid, glutamic acid via the amino acid, glutamine. Whatever may be its source, ornithine is decarboxylated by 'ornithine decarboxylase' of microbes like *E. coli* to produce putrescine and carbon dioxide. Putrescine is a diamine

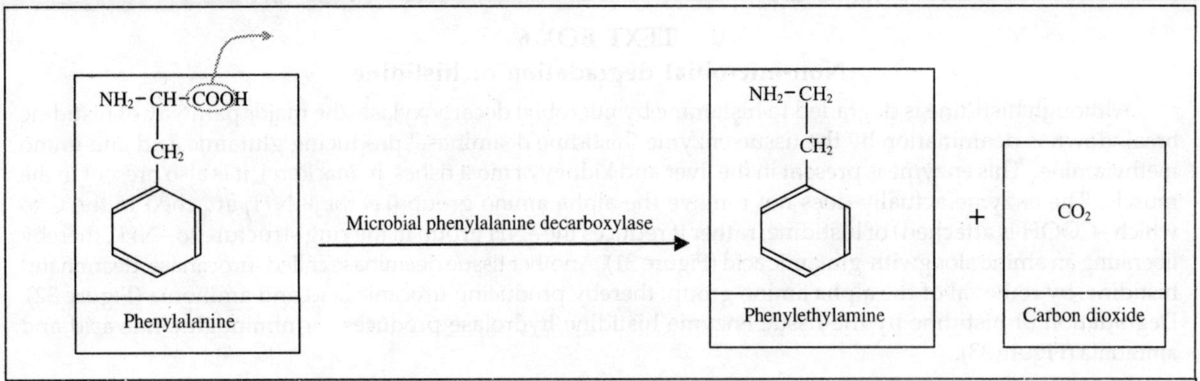

Figure 34: Microbial decarboxylation of phenylalanine

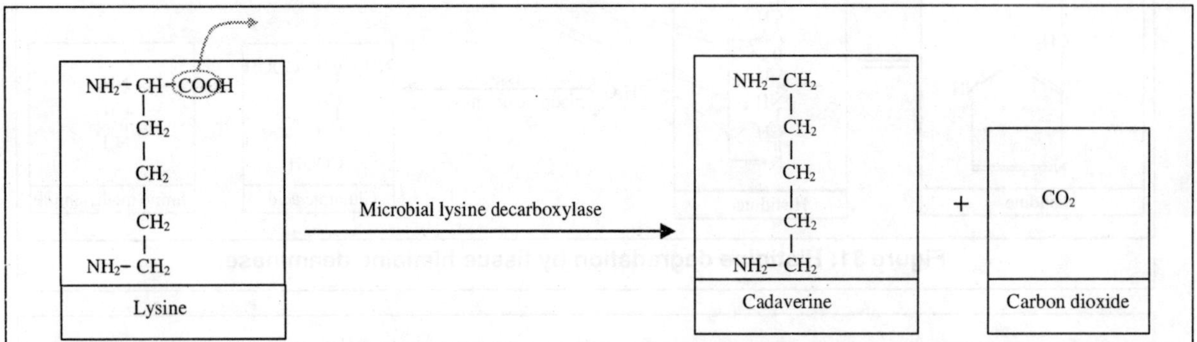

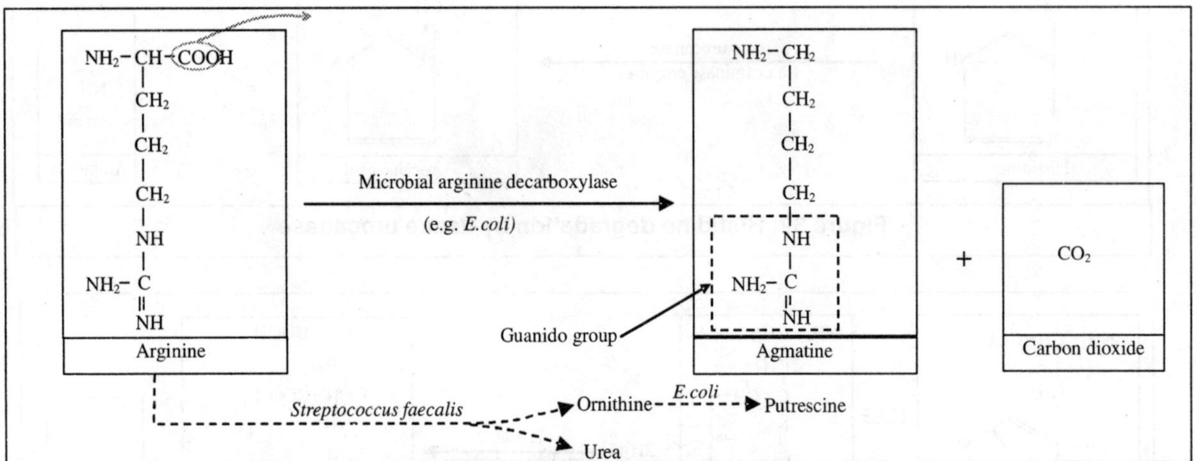

Figure 36: Microbial decarboxylation of arginine

and comes under ptomaines or biogenic amines. It is further converted to 'spermidine', which is ultimately converted to 'spermine'. However, the most important product of ornithine degradation is putrescine.

VI. Tyrosine

Tyrosine is decarboxylated by microbial 'tyrosine decarboxylase' to cresol and carbon dioxide (Figure 38). Sometimes it is also degraded by microbial 'tyrosine reductase' to phenol and alanine (Figure 39).

VII. Tryptophan

The amino acid tryptophan is deaminated by microbial deaminase called 'tryptophanase' to produce indole, pyruvic acid (a keto acid) and ammonia (Figure 40). Tryptophan is sometimes degraded in another way to produce skatole, glyoxylic acid and ammonia (Figure 41).

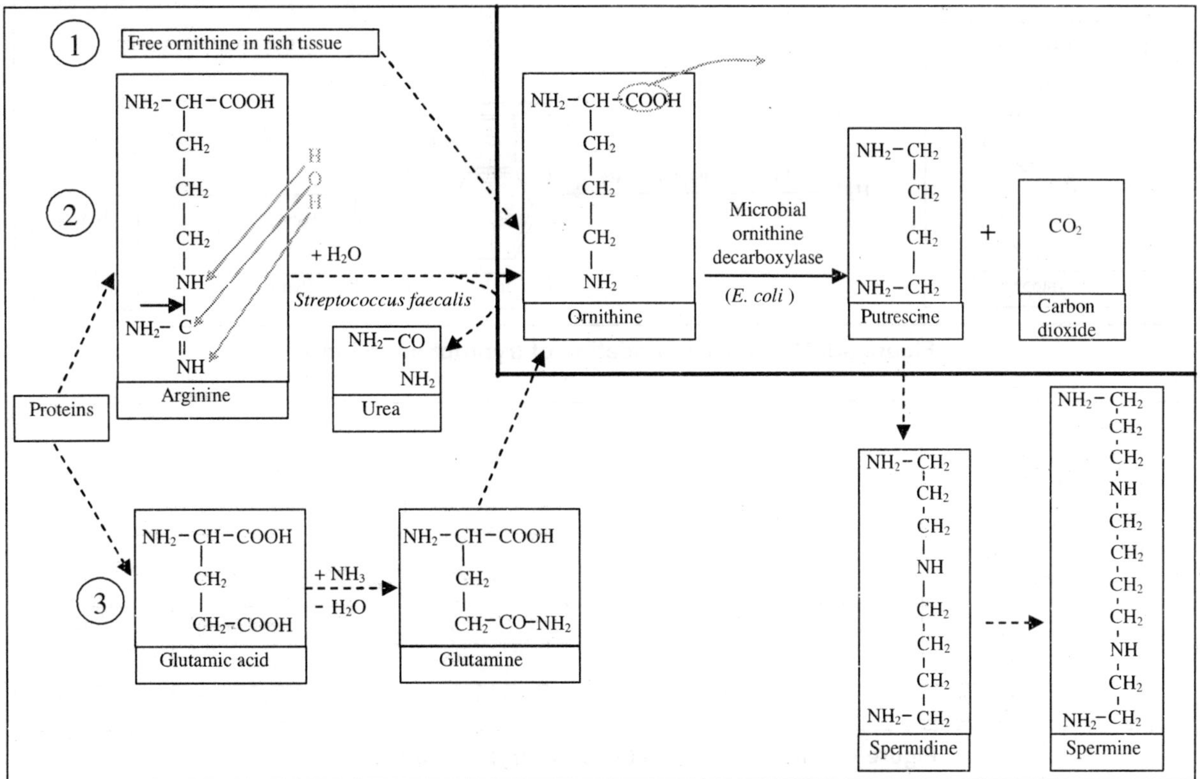

Figure 37: Microbial decarboxylation of ornithine

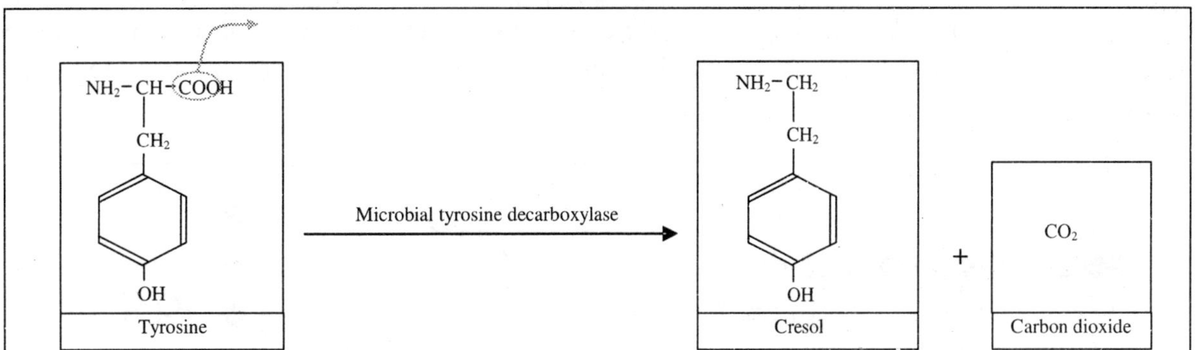

Figure 38: Microbial decarboxylation of tyrosine to cresol

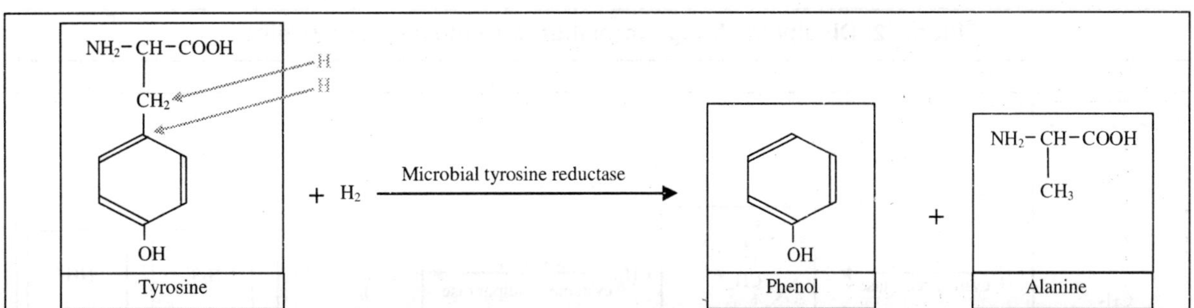

Figure 39: Microbial degradation of tyrosine to phenol

VIII. Cysteine

Cysteine and methionine are the two amino acids containing sulphur, which therefore produce sulphur-containing end products. Cysteine is usually present in the protein, forming bond between two polypeptide chains. Two cysteine molecules bind with each other at the sulphur ends forming a

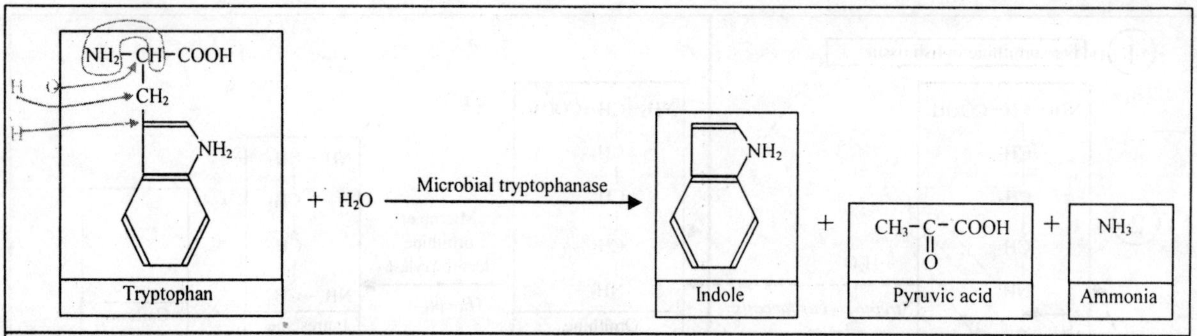

Figure 40: Microbial degradation of tryptophan to indole

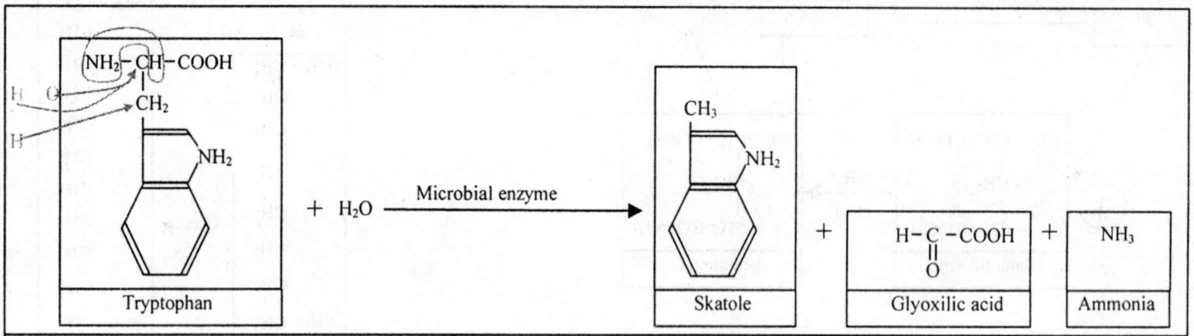

Figure 41: Microbial degradation of tryptophan to skatole

disulphide bridge, which is responsible for inter-polypeptide bonds. (Figure 42). The two cysteine molecules with disulphide bridge is called a 'cystine'[4]. During microbial putrefaction, cystine is reduced to form two molecules of cysteine. Cysteine is degraded by microbial 'cysteine desulphurase' to pyruvic acid, hydrogen sulphide and ammonia (Figure 43).

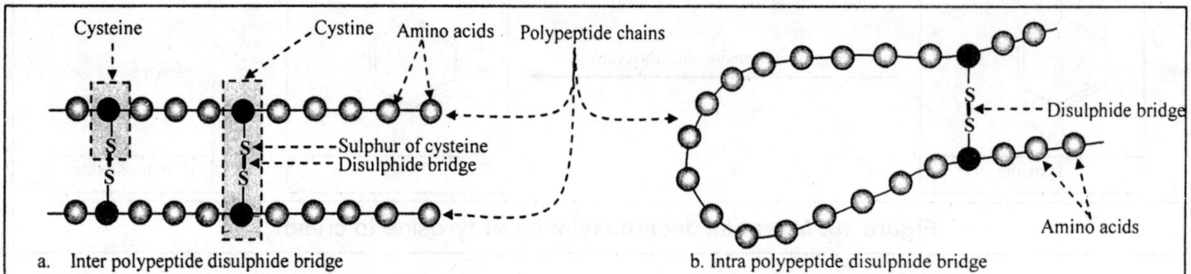

Figure 42: Disulphide bridge formation with the help of cysteine

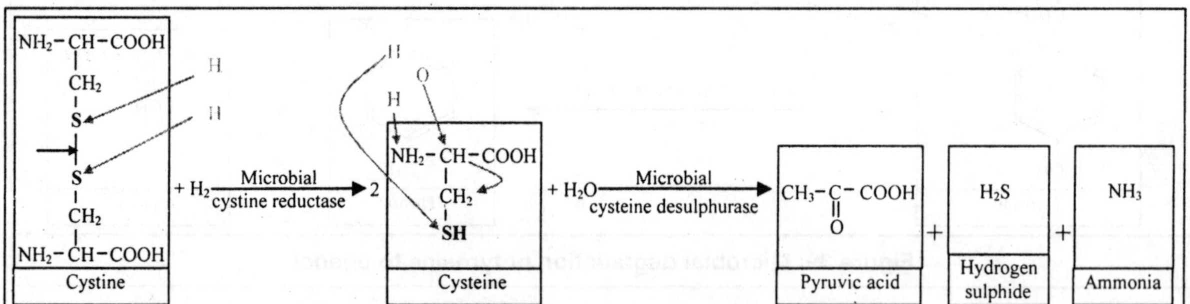

Figure 43: Microbial degradation of cysteine

4 Note –ei- in cyst<u>ei</u>ne as compared to –i- in cyst<u>i</u>ne to differentiate between them.

IX. Methionine

The sulphur-containing amino acid methionine is degraded by the microbial enzyme 'methionine reductase' to 2-amino butyric acid and methyl marcaptan (Figure 44).

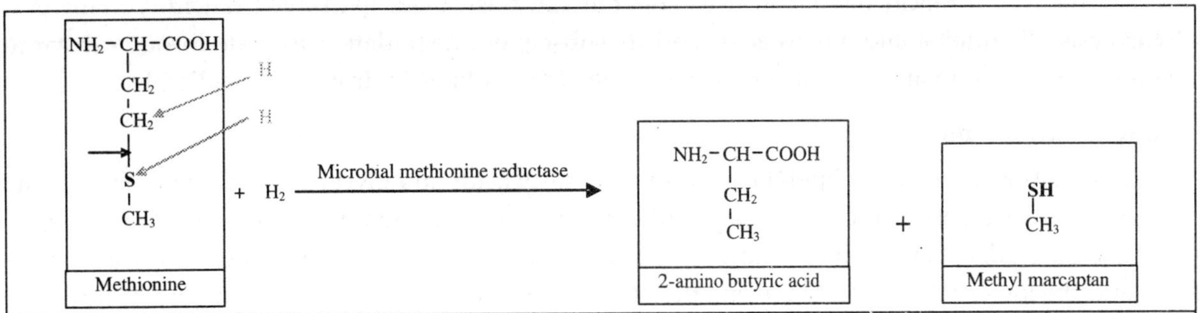

Figure 44: Microbial degradation of methionine

2. Hydrolysis of Complex Tissue Components into Simpler Substances and their Subsequent Utilisation

Microbes first utilise the simpler substances available freely in the fish tissue such as TMAO, urea and the free amino acids as well as more importantly those produced during autolysis such as the abundant amino acids and fatty acids (Figure 45). When these substances are almost completely exhausted, they start hydrolysing the complex tissue components such as proteins and lipids to produce simpler substances, which are utilised by them. However, the hydrolysis of proteins into amino acids and their utilisation by microbes with liberation of off-odour, off-flavour substances is more important with respect to fish spoilage than hydrolysis of lipids, particularly fats, into fatty acids and glycerol.

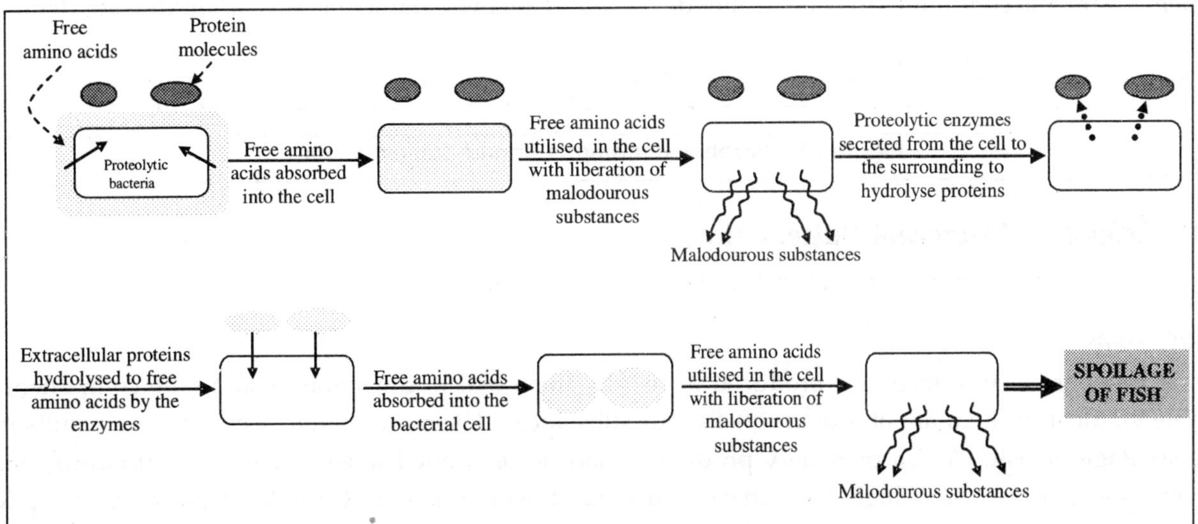

Figure 45: Hydrolysis of proteins by bacterial enzymes into amino acids and their subsequent degradation

For utilisation of proteins, the proteolytic[5] group of bacteria secret proteolytic enzymes from their cells to the surrounding. These enzymes hydrolyse the proteins into amino acids. Proteins cannot enter into the microbial cells because of their macromolecular size (large size). However, the building blocks of proteins are the amino acids, which are micromolecules (small size). They can enter into the microbial cells. Inside the microbial cells, they are metabolised and the end products are liberated to outside the cells. These end products are mostly off-odour, off-flavour substances presence of which in the tissue renders the fish spoiled.

5 Proteolytic bacteria are those bacteria, which can hydrolyse proteins.

The utilisation of lipids, particularly fats (triglycerides) involves lipolytic microbes[6]. They secret lipolytic enzymes (lipases) to the surrounding and hydrolyse fats into fatty acids and glycerol. Subsequently, they are utilised by microbes. The accumulation of free fatty acids in fish tissue leads to the development of a rancid flavour in fish. This rancidity[7] is called 'hydrolytic rancidity'. However, the hydrolysis of proteins into amino acids and its subsequent degradation to malodorous substances by the microbes is the main cause of spoilage and the effect of lipid hydrolysis is negligible.

3. Rancification Fats

Besides hydrolysis of fats by lipolytic microbes into fatty acids and glycerol, certain microbes contain the enzyme 'lipoxygenase' which activates oxidation of the unsaturated fatty acids of fish fats (Figure 46). The fats get oxidised in presence of atmospheric oxygen with the production of peroxides, aldehydes, ketones and the lower fatty acids. These end products are responsible for the undesirable rancid flavour of fish and the resulting rancidity is called 'oxidative rancidity'.

Fats + O_2 —(Microbial lipoxygenase)→ Peroxides, aldehydes, ketones and lower fatty acids ⟹ Oxidative rancidity

Figure 46: Rancification of fats by a microbes

The main component of the phospholipid lecithin is 'choline' which is decomposed by microbes to simple monoamines (mono-, di- and trimethylamine *i.e.* MA, DMA, TMA) as well as to the toxic oxyammonium compound, 'neurine'. Neurine is a ptomaine.

4. Nucleic Acid Decomposition

The nucleic acids (DNA and RNA) are decomposed by microbes to hypoxanthine and xanthine, which, in suitable conditions, may be further degraded by them to ammonia and carbon dioxide (Figure 47).

Nucleic acids (DNA and RNA) → Hypoxanthine and xanthine → NH_3 + CO_2

Figure 47: Microbial decomposition of nucleic acids

VII. Effects of Microbial Putrefaction

The microbial putrefaction of fish has the following effects.

1. Spoilage of Fish

A number of substances are produced in fish by microbial putrefaction of its tissue components (Table 9). Most of them are off-odour, off-flavour substances. Their cumulative accumulation results in the spoilage of fish. 'A fish or fishery product is said to be spoiled when it contains off-odour, off-flavour substances at such high concentration that the consumer rejects it due to its poor organoleptic qualities'[8].

---

6 Lipolytic microbes are those microbes, which can hydrolyse fats into glycerol and fatty acids.

7 Rancidity is the off-flavour produced due to deterioration of lipids or fats. It is of two types.

1. Oxidative rancidity, which is caused by the oxidation of the unsaturated fatty acids of tissue fat by atmospheric oxygen and
2. Hydrolytic rancidity, which is caused by the accumulation of the free fatty acids produced by hydrolysis of fats

Between the two, oxidative rancidity is more important for the development of rancidity in fish and fishery products.

8 Organoleptic qualities or sensory qualities are those qualities of fish or any other food products, which can be sensed by human sense organs. These include taste, flavour, odour, appearance and texture.

Table 9: Products of microbial putrefaction of fish

1.	**Common gases with no off-odour**	5.	**Volatile aromatic alcohols**
	a. Hydrogen		a. Phenol
	b. Carbon dioxide		b. Cresol
2.	**Volatile base nitrogenous substances**	6.	**Heterocyclic compounds**
	a. Ammonia		a. Indole
	b. Simple monoamines		b. Skatole
	i. Trimethylamine (TMA)	7.	**Aromatic acids**
	ii. Dimethylamine (DMA)		a. Benzoic acid
	iii. Methylamine (MA)		b. Phenylpropionic acid
3.	**Biogenic amines (ptomaines)**	8.	**Ammonium salts of aromatic acids**
	a. Cyclic monoamines		a. Ammonium benzoate
	i. Histamine		b. Phenyl ammonium propionate
	ii. Phenylethyamine	9.	**Lower fatty acids**
	b. Non-cyclic diamines		a. Acetic acid
	i. Cadaverine		b. Propionic acid
	ii. Putrescine		c. Valeric acids
	iii. Agmatine		d. Lactic acid
	c. Oxyammonium base		e. Succinic acid
	i. Neurine	10.	**Lipid oxidation products** (microbial rancification)
4.	**Volatile sulphur compounds**		a. Peroxides
	a. Hydrogen sulphide		b. Aldehydes
	b. Marcaptans		c. Ketones
	i. Methylmarcaptans		
	ii. Ethylmarcaptans		

The important substances responsible for spoilage of fish are given below.

1. *Volatile Base Nitrogenous Substances*: TMA, DMA, MA and ammonia
2. *Volatile Sulphur Compounds*: Hydrogen sulphide and marcaptans
3. *Volatile Aromatic Alcohols*: Phenol and cresol
4. *Heterocyclic Compounds*: Indole and skatole

Most important among these substances are TMA and ammonia. Though all fishes undergo spoilage if kept unpreserved, their pattern of spoilage mainly depends upon their habitat as described below.

A. Freshwater Fishes

Among the volatile bases, which are the main substances responsible for spoilage, 'ammonia' is most important in freshwater fishes. Here ammonia is produced through deamination of amino acids.

B. Marine Teleosts

In marine teleosts, the most important volatile base is 'TMA'. It is produced in substantial quantities through reduction of TMAO, which is an osmo-regulatory substance present in marine fishes.

C. Marine Elasmobranchs

In marine elasmobranchs (cartilaginous fishes), the most important volatile base is 'ammonia'. It is produced through decomposition of urea. Elasmobranchs contain high amount of urea in their tissue. However, TMA is also produced from TMAO in the elasmobranchs, but not as substantially as in marine teleosts.

2. Increase in pH

After death, the pH of fish tissue falls from neutral (pH $\approx$ 7.0) to acidic range (pH 6.2–6.5) due to accumulation of lactic acid produced by anaerobic oxidation of tissue glycogen (Figure 48). The tissue pH remains acidic till microbial spoilage sets in. The microbial decomposition of fish tissue results in the production of large quantities of nitrogenous substances (ammonia, TMA, DMA, MA, mono- and di-amines), which are basic in nature. As they are basic they are called nitrogenous bases. The accumulation of these bases increases the pH of fish tissue from acidic to alkaline range (pH 7.1–7.2) after neutralising the acidity.

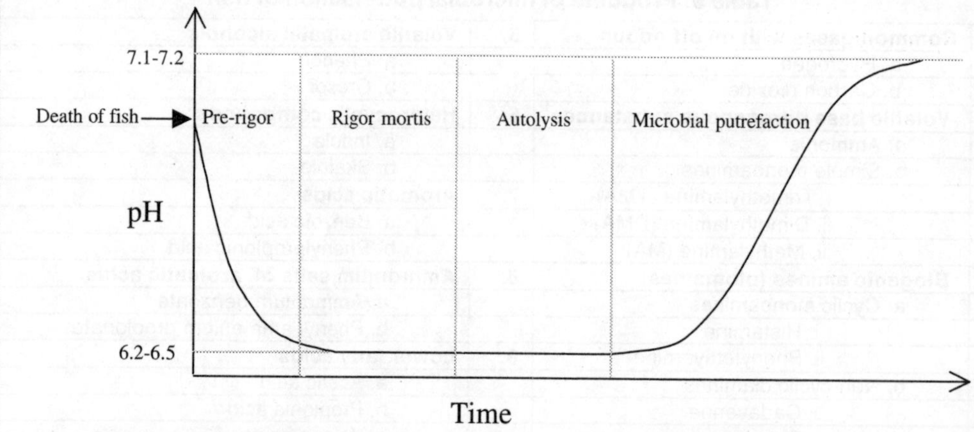

Figure 48: Increase in pH of fish tissue due to microbial putrefaction

3. Increase in NPN Substances

The non-protein nitrogenous (NPN) substances such as TMAO, urea and free amino acids are the most favoured nutrients for microbial growth. Small quantity of NPN substances is present in live fish. After death, the process of autolysis hydrolyses the tissue proteins producing substantial quantity of free amino acids thereby drastically increasing the NPN content of fish tissue. High content of NPN favours microbial growth. During microbial putrefaction too, the proteolytic microbes secret proteolytic enzymes to outside their cells to hydrolyse the surrounding tissue proteins into amino acids. This results in a further increase in NPN of fish tissue. The increased NPN is not affected by microbial decomposition, though microbes decompose them. This is because; microbes decompose them into some other substances, which retain the nitrogen present in the original NPN and the product NPN accumulates in fish tissue. The loss of nitrogen through volatilisation of the volatile base nitrogenous (VBN) substances (TMA, DMA, MA, ammonia and amines) is negligible.

4. Development of Rancidity

During microbial putrefaction, the lipolytic microbes secret lipolytic enzymes to the surrounding, which hydrolyse the tissue fats into glycerol and fatty acids. Accumulation of these free fatty acids leads to the development of a type of rancidity called 'hydrolytic rancidity'. Some microbes contain the enzyme lipoxygenase (lipoxidase) which catalyses the oxidation of the unsaturated fatty acids of tissue fats by atmospheric oxygen. The end products are peroxides, hydroperoxides, aldehydes, ketones and lower fatty acids. The presence of these oxidation products leads to the development of a type of rancidity called 'oxidative rancidity'. Though microbial activity in fish tissue results in the development of hydrolytic and oxidative rancidity, their contribution to the rancidity in fish tissue is insignificant. The hydrolytic rancidity that develops due to hydrolysis of tissue fats by the endogenous tissue enzymes during autolysis and the oxidative rancidity that develops due to autoxidation of tissue lipids are the major cause of development of rancidity in fish and fishery products.

5. Food Poisoning

Some of the end products of microbial putrefaction are toxic and can cause food poisoning. They include histamine, phenylethylamine, cadaverine, putrescine, neurine, phenol, cresol, indole and skatole.

VIII. Control of Microbial Putrefaction

Microbes are the main agent of fish spoilage. Control of microbial putrefaction not only helps in keeping fish fresh for long time, it also helps in preserving fish over prolonged duration depending on the method of preservation adopted. Some of the methods adopted to control microbial putrefaction are given below.

1. Washing the Fish

Immediately after fish is removed from water, it should be thoroughly washed with potable water so as to remove the adhering dirt and mud which harbour high load of spoilage microbes.

2. Gutting and Gilling

In case of large size fishes, the gut and gills, which have high load of natural microbial flora, are removed immediately after harvesting. Care is taken not to puncture the gut; otherwise the gut microbes would contaminate the sterile muscle. Sometimes scales as well as skin are removed to eradicate the skin microbes. However, such practice is not practicable in case of small fishes caught in bulk quantities. In such case, gutting, gilling and scaling is done by the end user instead of being done by the fisherman.

3. Maintenance of Proper Hygiene during Post Harvest Handling and Processing

Besides the microbial flora already present in fish when harvested, spoilage microbes may also get access to fish body during post harvest handling and processing. These microbes are called the 'contaminating microbes'[9]. The sources of these microbes are the uncleaned or improperly cleaned processing equipments, processing tables, knives, dressing boards, hands of processing personnel, dirty ice, dirty water and other such dirty materials which come in contact with fish. The load of these contaminating microbes can be drastically reduced by maintenance of proper hygiene. It includes proper cleaning of fish using clean potable water after it enters into the processing hall as well as after its dressing. It also includes thorough cleaning of the surfaces coming in contact with fish, use of potable water, use of uncontaminated ice prepared from potable water, avoiding use of dirty materials, preventing access of contaminating agents such as flies, rodents and other pests as well as adoption of proper personnel hygiene of the processing persons. Personnel hygiene includes proper cleaning of hands and feet before entering into the processing hall, using aprons, head and mouth covers etc.

4. Chilling

Immediately after harvesting and quickly following some preliminary operations like washing, gutting and gilling, fish should be chilled rapidly by icing, chilled seawater stowage (CSW) or refrigerated seawater stowage (RSW). In chilling of fish, the temperature of fish body is lowered to a temperature of about 0°C, which not only retards microbial activity considerably, but also slows down other spoilage processes remarkably. Chilling can keep fish in good condition for few days (usually a week or so).

5. Freezing

In freezing of fish, the temperature of fish body is lowered to a very low temperature, much below the freezing point of fish, by removal of heat from its body using artificial refrigeration system. Usually during freezing of fish, its temperature is lowered to about –35°C to –40°C at which the fish body solidifies and becomes hard. Following the freezing operation, the frozen fish is stored at a very low temperature of less than –18°C in a cold store. Unlike chilling which 'slows down' microbial activity, the very low temperature of freezing and cold storage almost 'completely arrest' it. This is because, below –10°C microbial activity is completely arrested. Freezing can keep fish in good condition for few months (usually 3-6 months).

6. Pasteurisation

In preservation of fish or fishery product by pasteurisation, they are exposed to moist heat reaching a maximum temperature of 100°C, the boiling point of water. Processing is done by boiling in open hot water baths or by steaming. At such temperatures, most of the non-spore formers die while the spore-formers survive by forming heat resistant spores. Pasteurisation is adopted in the processing of fish sausage, fish ham, fish balls, fish cakes, kamaboko, sous vide etc. Milk is the most important food item to be processed by pasteurisation. Pasteurisation can keep these fishery products in good condition for 2-3 days as the spore-formers, which survive the pasteurisation heat treatment, grow and spoil the

---

9 During post harvest handling and processing of fish, there is great risk of contamination with both spoilage and pathogenic microbes. Post harvest contamination is the main source of pathogenic microbes into fish and fishery products, because most of the natural microbial flora of fish and its environment are spoilage type and microbes pathogenic to human being are found in materials and environment coming in contact with human activity.

products during storage beyond this period. Storage of pasteurised products under cooler temperatures can extend their shelf life further to a maximum period of a week or so.

7. Sterilisation

In preservation of fish or fishery product by sterilisation, they are exposed to moist heat above 100°C, the boiling point of water. Such heat sterilisation is done by pressure processing in closed retorts. At such high temperatures, all the non-spore formers as well as the heat resistant spore-formers of significance die. Sterilisation is adopted in the processing of fish and fishery products, which are to be kept for months together under normal non-refrigerated condition of storage and distribution. Such products include canned fish and retorted sausage. These products are heat sterilised to attain 'commercial sterility'. A product is said to have attained commercial sterility, if it neither spoils nor causes any health hazard in the consumer during normal non-refrigerated condition of storage and distribution. That means, the products are not completely free of microbes and therefore have not attained 'absolute sterility', rather some microbes survive the heat treatment, but the condition of the can (pH of the food, storage temperature) does not allow them to grow and cause spoilage or cause health hazard by producing toxins.

8. Reduction of Moisture

All microbes need water or moisture for their growth and activity. The main constituent of fish body is water contributing to about 70-80 per cent of body weight. Out of the rest, the most important component is protein, which is an excellent nutrient for most microbes. Thus fish body is a nutrient-rich, moisture-rich medium for microbial growth and activity. Therefore, reduction of moisture content of fish body brings about reduction of microbial activity thereby preventing spoilage. Reduction of moisture is the basis of preservation of fish by drying, salting and smoking. The shelf life of these products varies depending on the residual moisture in them. Generally, they remain good for 2-3 months at ambient conditions.

9. Acidification

Most of the microbes causing fish spoilage need near-neutral condition for their activity. Acidic conditions drastically retard their activity. In preservation of fish by acidification, the fish is acidified by addition of acids to such an extent that activity of most of the spoilage microbes is almost completely inhibited. Fish preservation by acidification includes marinading and pickling. In marinading, fish is mixed with table salt and vinegar. The acetic acid in vinegar brings about the acidification. In pickling process adopted to prepare fish pickle and prawn pickle, the ingredients such as vinegar, tamarind or lemon acidify the product by the acetic acid, tartaric acid and citric acid they contain respectively.

10. Fermentation

Fermentation is the process of break down of organic substances by enzymes under anaerobic condition. Initially assumed to be a microbial process by Cagniard-Latour, Schwann and Pasteur, later it was confirmed by Buchner to be an enzymatic process. Here mostly microbial enzymes are involved. He showed that a cell-free extract obtained by filtering yeast that had been ground with quartz and sand and that contained no yeast cell could still ferment sugar to alcohol and CO_2 (Figure 49).

$$\text{Sugar} \xrightarrow{\text{Microbial enzyme in cell-free extract}} CO_2 + \text{Alcohol}$$

Figure 49: Fermentation of sugar

Preservation of fish by fermentation is of two types as given below.

(*a*) Salt Fermentation

Here fish is mixed with high proportion of salt (20-25 per cent) and stored closed in a container for months together. The high salt content prevents microbial growth by acting as a preservative. Some of the tissue enzymes which do not denature even at the very high concentration of salt continue to hydrolyse the tissue components due to which very often the product becomes liquid. The enzymes also ferment (decompose) the tissue components to simpler substances, which add flavour to the products. Thus, while enzyme activity continues unabated, the salt prevents microbial spoilage of fish by inhibiting microbial break down of these simpler substances into off-odour, off-flavour substances. Salt fermented products include fish sauce, nuoc-mam, fish patis, bagoong, Gaffelbidder, herring tit-bids etc.

(*b*) Microbial Fermentation

In preservation of fish by microbial fermentation, growth and activity of spoilage microbes is inhibited by lowering of pH of fish tissue by the fermentating microbes, which can produce acids from extraneous substrates such as sugars added to fish (Figure 50). The fermentation takes place either by the endogenous microbes (present in the fish body) or by an inoculum of exogenous microbes (added to fish). For example, by adding glucose as substrate and lactic acid bacteria cultures (LAB cultures *i.e.* cultures of *Lactobacillus sp.*) as inoculated fermentators to fish as well as by providing anaerobic condition, glucose is converted to lactic acid by the LAB. The acid reduces the pH and inhibits the growth of other undesirable microbes thereby preserving the fish. Products prepared by microbial fermentation include fermented sausage, i-sushi, koji-juki, katsuobushi (a product fermented by mold) etc.

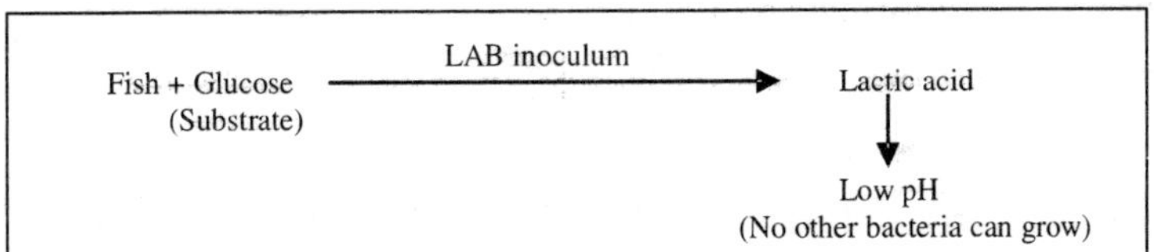

Figure 50: Microbial fermentation of fish

11. Irradiation

In irradiation preservation of fish, it is exposed to an 'ionising radiation'[10], which kills most of the microbes present in it. Thus these radiations prevent microbial spoilage of fish. The use of γ-rays produced from ^{60}Co and the X-rays have been experimentally tried for preservation of fish.

12. Use of Chemicals

A number of chemicals are used to retard or prevent microbial spoilage of fish and fishery products. Such chemicals are usually called 'preservatives'. They include the following.

(*i*) Ammonia

(*ii*) Chlorine

(*iii*) Sulphur dioxide

(*iv*) Acids: Formic acid, acetic acid, propionic acid, benzoic acid and sorbic acid

---

10 **Ionising radiation:** Energy transmitted through 'space' in a variety of forms is generally called 'radiation'. Most important among the various types of radiations is the 'electromagnetic radiation', which includes light rays and X-rays. Among the electromagnetic radiations, those which have energy sufficiently high (more than 10eV) to ionise cell components so that they can no longer perform critical cellular functions and consequently damage the cells are called 'ionising radiations'. The ionising radiations are produced by formation of charged subatomic particles (electrons, protons, neutrons) from atoms or molecules. These radiations have lethal effect on microbial cells. The different types of ionising radiations are as follows.

α-rays	Produced during disintegration of nuclei of radioactive elements like ^{60}Co, ^{90}Sr, ^{127}Cs etc.
β-rays	
γ-rays	
X-rays	Produced by powerful electrical accelerators
High speed electron beam	

(*v*) Salts of acids

Acid	Salts of acid
Formic acid	Sodium formate, Potassium formate, Calcium formate
Acetic acid	Sodium acetate, Potassium acetate, Calcium acetate, Sodium diacetate
Propionic acid	Sodium propionate
Benzoic acid	Sodium benzoate
Sorbic acid	Potassium sorbate, Sodium sorbate

(*vi*) Sulphites: Sodium sulphite, potassium sulphite, sodium bisulphite, potassium bisulphite, sodium metabisulphite, potassium metabisulphite

(*vii*) Nitrates: Sodium nitrate, potassium nitrate

(*viii*) Nitrites: Sodium nitrite, potassium nitrite

(*ix*) Hexamethylene tetramine

(*x*) Esters of parahydroxy benzoic acid

(*xi*) Hydrogen peroxide

(*xii*) Phosphate peroxide: Sodium pyrophosphate peroxide, potassium pyrophosphate peroxide, disodium hydrogen phosphate peroxide, dipotassium hydrogen phosphate peroxide

(*xiii*) 5-aminohexahydropyrimidines

(*xiv*) Tert-butyl hydroperoxide

(*xv*) Antibiotics:

- (*a*) Tetracycline: Oxytetracycline (OTC) and chlortetracycline (CTC)
- (*b*) Subtilin
- (*c*) Tylosin

(*xvi*) Bacteriocins

- (*a*) Nisin
- (*b*) Colicins
- (*c*) Subtilisin

13. Modified Atmosphere Packaging

Modified atmosphere packaging (MAP) is used to extend the shelf life of fresh fish during chilled storage. In this method, fresh fish is packed in an airtight container within which the atmosphere is modified desirably by flushing-in of required gases in required proportion. The three major gases used commercially are CO_2, N_2 and O_2.The extension of shelf life in MAP is the result of the antimicrobial activity of these gases. CO_2 has bacteriostatic effect on microorganisms, N_2 inhibits the growth of aerobic microorganisms and O_2 inhibits the growth of strictly anaerobic microorganisms. In case of lean fishes all the three gases are used in the ratio of 40 per cent: 30 per cent: 30 per cent:: CO_2: N_2:O_2. But in case of fatty fishes, O_2 is not used as it oxidises the tissue fats and turns them rancid. Here, the gas mixture consists of CO_2 and N_2 in the ratio of 40-60 per cent and 60-40 per cent respectively.

IX. Summary

1. What is Microbial Putrefaction ?

'Microbial putrefaction' or 'microbial decomposition' is the process of conversion of fish tissue components by microorganisms into off-odour, off-flavour substances accumulation of which distracts the consumers and thereby results in the spoilage of fish.

2. Importance of Microbes in Fish Processing

Microbes are of great concern in fish processing due to the following reasons.

(*a*) Small Generation Time

Most of the microorganisms have extremely small generation time.

(*b*) Resistance

There are few microbes, which are resistant to extremely unfavourable conditions.

(*c*) Invisibility

Since they cannot be seen with naked eye, their presence is very often underestimated.

(*d*) Agent of Fish Spoilage

They are the most important agent of fish spoilage.

(*e*) Agent of Health Hazards

Some microbes present in fish cause health hazards in persons who consume it.

3. Types of Microbes in Fish

The microbes associated with fish spoilage and health hazards are of two types as follows.

A. Endogenous Microbes

Endogenous microbes, also called 'natural microbial flora' or 'resident microbes' are those microbes, which inhabit the fish body when fish is alive, but do not cause any disease in it. These microbes inhabit those parts of the fish body, which are exposed to the surrounding (Figure 20). Microbes inhabit the skin, gills and gut in the increasing order of their load while muscle and internal organs are sterile (free of microbes) in healthy fish.

B. Exogenous Microbes

They are the microbes, which get their access to the fish body during post harvest handling, processing and storage. Their main sources are the twines of the nets, mud adhering to fish body and nets, the decks, the pound boards of fish-holds used for storing fish in fishing vessels, cutting boards, washing tubs, water used for washing fish, ice used for chilling fish, floor of the processing halls, processing tables, knives, utensils and above all the hands of the processors.

4. Species of Microbes Associated with Spoilage of Fish

It is important to note that generally the endogenous microbes or the natural microbial flora present in fish are spoilage type while the exogenous microbes contaminating the fish during post harvest handling, processing and storage are pathogenic type. The different species of microbes associated with the spoilage of fish through microbial putrefaction are *Pseudomonas, Aeromonas, Achromobacter, Flavobacterium, Vibrio, Moraxella, Arthrobacter, Micrococcus, Corynebacterium, Photobacterium, Bacillus, Shewanella* (formerly *Altromonas*) and *Proteus.*

5. Spread of Microbes in the Body of Dead Fish

After death of fish, microbes penetrate into the muscle and internal organs due to the following reasons.

(*a*) Loss of defense mechanism or immunity of body.

(*b*) Loss of microbe-impermeability of the membranous linings of the body.

(*c*) Abundance of readily utilisable nutrients for microbial growth and multiplication.

6. Mechanism of Microbial Putrefaction

Microbes decompose and spoil fish in four ways as given below.

I. Utilisation of readily utilisable simple substances such as NPN (non-protein nitrogenous) substances

II. Hydrolysis of complex tissue components (*e.g.* proteins) into simpler substances (*e.g.* amino acids) and their subsequent utilisation.

III. Rancification

IV. Nucleic acid decomposition

I. Utilisation of Readily Utilisable Simple Substances such as NPN

The first substances to be utilised by the microbes are the readily utilisable simple substances, mainly the NPN components. They include TMAO, urea and amino acids.

A. *TMAO Degradation*

TMAO is sequentially degraded as given below (Also see Figure 21).

Trimethylamine oxide (TMAO) → Trimethylamine (TMA) → Dimethylamine (DMA) → Methyl amine (MA) → Ammonia (A)

Among the break down products of TMAO, TMA is most important, as it is responsible for the typical fishy odour of marine teleosts. In the gadoid fishes like Alaska pollack, cod and haddock, TMAO is degraded to DMA and formaldehyde. The formaldehyde deteriorates protein quality of fish during frozen storage (Figure 24).

B. *Urea Degradation*

Urea present in the tissues of cartilaginous fishes is decomposed to carbon dioxide (CO_2) and ammonia (NH_3) by the microbial enzyme 'urease' as given below. The ammonia imparts the ammoniacal odour of spoiling cartilaginous fishes (Also see Figure 25).

Urea → Carbon dioxide + Ammonia

C. *Free Amino Acid Degradation*

The free amino acids present in fish and those produced after death are absorbed into the microbial cells where they are metabolised to intense off-odour, off-flavour substances accumulation of which in the fish tissue renders the fish spoiled.

The decomposition of important amino acids by microbes is given below.

1. Histidine → Histamine + CO_2 (Figure 29)
2. Phenylalanine → Phenylethylamine + CO_2 (Figure 34)
3. Lysine → Cadaverine + CO_2 (Figure 35)
4. Arginine → Agmatine + CO_2 (Figure 36)
5. Ornithine → Putrescine + CO_2 (Figure 37)
6. Tyrosine → Cresol + CO_2 (Figure 38) OR Phenol + Alanine (Figure 39)
7. Tryptophan → Indole + NH_3 (Figure 40) OR Skatole + 2-keto acetic acid + NH_3 (Figure 41)
8. Cysteine → Pyruvic acid + H_2S + NH_3 (Figure 43)
9. Methionine → Methyl marcaptan + 2-amino butyric acid (Figure 44)

The presence of histamine in foods in high concentration causes a type of food poisoning called 'histamine poisoning' or 'histamine toxicity' or 'scombroid poisoning' as it is mostly caused by fishes belonging to the family 'scombridae', which have high histidine content. The most important microbe causing histamine production is *Morganella morganii* (previously *Proteus morganii*). Histamine production in the fish tissue can be reduced by (1) Quick chilling of fish immediately after harvesting by icing, chilled seawater stowage (CSW) or refrigerated seawater stowage (RSW), (2) Immediate freezing of fish, (3) Maintenance of good hygienic practices and (4) Processing fish as quickly as possible.

II. Hydrolysis of Complex Tissue Components into Simpler Substances and their Subsequent Utilisation

Microbes first utilise the simpler substances available freely in the fish tissue such as TMAO, urea and the free amino acids as well as more importantly those produced during autolysis such as the abundant amino acids and fatty acids (Figure 45). When these substances are almost completely exhausted, they start hydrolysing the complex tissue components such as proteins and lipids to produce simpler substances to be utilised by them.

III. Rancification Fats

Certain microbes contain the enzyme 'lipoxygenase' which activates oxidation of the unsaturated fatty acids of fish fats (Figure 46). The fats get oxidised in presence of atmospheric oxygen with the production of peroxides, aldehydes, ketones and the lower fatty acids. These end products are responsible for the undesirable rancid flavour of fish and the resulting rancidity is called 'oxidative rancidity'.

IV. Nucleic Acid Decomposition

The nucleic acids (DNA and RNA) are decomposed by microbes to hypoxanthine and xanthine which, in suitable conditions, may be further degraded by them to ammonia and carbon dioxide (Figure 47).

7. Effects of Microbial Putrefaction

(*a*) Spoilage of Fish

A number of off-odour, off-flavour substances are produced in fish by microbial putrefaction of its tissue components. Their cumulative accumulation results in the spoilage of fish. Some of these substances include NH_3, H_2S, TMA, indole, skatole etc.

(*b*) Increase in pH

After death, the pH of fish tissue falls from neutral (pH $\approx$ 7.0) to acidic range (pH 6.2–6.5) due to accumulation of lactic acid (Figure 48). The microbial decomposition of fish tissue results in the production of large quantities of nitrogenous substances, which are basic in nature. The accumulation of these bases increases the pH of fish tissue from acidic to alkaline range (pH 7.1- 7.2).

(*c*) Increase in NPN Substances

During microbial putrefaction, the proteolytic microbes secret proteolytic enzymes to outside their cells to hydrolyse the surrounding tissue proteins into amino acids. This results in an increase in NPN of fish tissue.

(*d*) Development of Rancidity

During microbial putrefaction, the lipolytic microbes hydrolyse the tissue fats into glycerol and fatty acids. Accumulation of these free fatty acids leads to the development of 'hydrolytic rancidity'. Some of the microbes, which contain the enzyme lipoxygenase (lipoxidase) catalyse the oxidation of the unsaturated fatty acids of tissue fats by atmospheric oxygen. The end products are peroxides, hydroperoxides, aldehydes, ketones and lower fatty acids. The presence of these oxidation products leads to the development of 'oxidative rancidity'.

(*e*) Food Poisoning

Some of the end products of microbial putrefaction are toxic and can cause food poisoning. They include histamine, phenylethylamine, cadaverine, putrescine, neurine, phenol, cresol, indole and skatole.

8. Control of Microbial Putrefaction

Some of the methods adopted to control microbial putrefaction are (*a*) Washing the fish, (*b*) Gutting and gilling, (*c*) Maintenance of proper hygiene during post harvest handling and processing, (*d*) Chilling, (*e*) Freezing, (*f*) Pasteurisation, (*g*) Sterilisation, (*h*) Reduction of moisture, (*i*) Acidification, (*j*) Fermentation, (*k*) Irradiation, (*l*) Use of chemicals and (*m*) Modified atmosphere packaging.

6

Lipid Oxidation (Autoxidation)

I. What is Lipid Oxidation (Autoxidation) ?

Fish lipids are rich in unsaturated fatty acids. These fatty acids are mostly present in the triglyceride molecules except when the hydrolysis of the triglycerides makes them free to form the free fatty acids (FFA). The unsaturation of the fatty acids, whether free or in the triglyceride molecules, is due to the presence of double bonds between carbon atoms in their hydrocarbon chains (Figure 51). These points of unsaturation are greatly susceptible to attack by atmospheric oxygen. When exposed to atmosphere, the atmospheric oxygen attacks and reacts with the double bonds on the fatty acids. The reaction of the atmospheric oxygen with the unsaturated double bonds of the fatty acids of lipid molecules is called 'lipid oxidation' or 'autoxidation'. Autoxidation is not just simply addition of oxygen to the lipid molecules at the points of unsaturation; rather it is the complete set of reactions involving addition of oxygen to the lipid molecules as well as subsequent reactions. Lipid oxidation results in the formation of different types of end products, the presence of which in fish and fishery products imparts a typical rancid flavour called 'oxidative rancidity'.

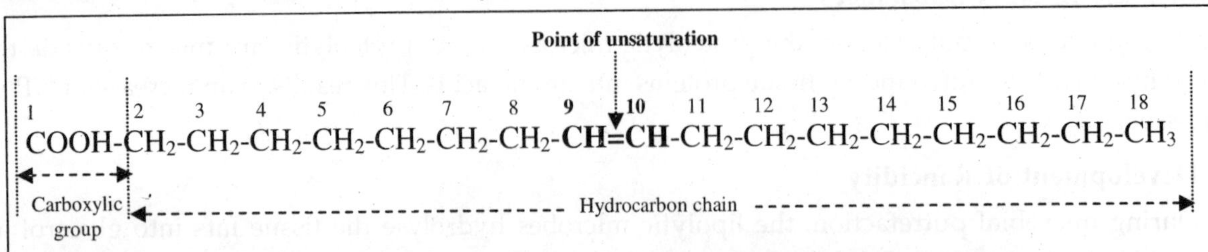

Figure 51: Structure of a fatty acid (oleic acid which has 18 carbon atoms and only one double bond between the carbon atoms 9 and 10)

When a fish is alive, its tissue lipids come in indirect contact with the atmospheric oxygen as it is transported to all the cells via blood circulation. This oxygen cannot oxidise the tissue lipids, because fish tissue, like tissue of most other organisms, contains the natural antioxidant tocopherol (Vitamin E), which is also circulated to the tissue via blood circulation. Natural antioxidants are substances, which inhibit lipid oxidation and occur naturally in plants and animals. Thus, the natural antioxidant tocopherol prevents lipid oxidation in live fish. Moreover, the living cells possess several protection mechanisms directed against lipid oxidation products. An enzyme, 'glutathione peroxidase' reduces the hydroperoxides produced during lipid oxidation to corresponding hydroxy- compounds. This reaction needs supply of reduced glutathione. After death of fish, blood circulation stops and therefore, tocopherol supply to the tissues ceases. As a result, the tissue lipids become liable to oxidation after death. At the same time, glutathione gets depleted in the cells due to which the lipid oxidation products accumulate in the tissue. Thus, lipid oxidation and accumulation of the lipid oxidation products start after death of fish and continue unabated. However, the rate of lipid oxidation completely depends upon the storage condition, particularly access of oxygen to fish tissue and storage temperature.

It is, however, important to note that, in case of fish to be marketed fresh (usually called wet fish), the microbial and enzymatic activities deteriorate and spoil the fish at such a rapid rate and to such an extent that the loss of quality due to autoxidation becomes insignificant as compared to that caused by the microbes and enzymes. Wet fish spoils before oxidative rancidity reaches any objectionable level. But in fishery products, particularly in frozen fish, in which microbial and enzymatic activities are almost completely inhibited, autoxidation continues unabated during cold storage, though at a slower rate, and dominates the loss of quality of frozen fish. Therefore, autoxidation is the dominant agent of deterioration in frozen fish whereas microbial and enzymatic deterioration dominate in wet fish. The same is also true for other fishery products like salted fish, dried fish and fishery byproducts (fish oils and fishmeal). Therefore, in this chapter autoxidation has been described not only as a post mortem change but also more importantly as an important agent of deterioration for other processed fishery products and byproducts.

II. Mechanisms of Lipid Oxidation

Lipids, when exposed to atmosphere, react with the atmospheric oxygen and get oxidised. Lipid oxidation may occur in two ways, (1) Enzymatically by the enzyme lipoxygenase (formerly called lipoxidase) or (2) Non-enzymatically by autoxidation. Lipoxygenase is present in fish and can oxidise lipids. However, non-enzymatic oxidation through autoxidation is more common than enzymatic oxidation. Therefore, autoxidation is of much greater concern in fish processing than enzymatic lipid oxidation. That is why, the terms 'lipid oxidation' and 'autoxidation' are used synonymously.

Lipid oxidation, which takes place through a non-enzymatic self-oxidation process, is called 'autoxidation' (*auto*: self; *oxidation*: reaction with oxygen). In autoxidation of lipids, the help of no other catalytic agent is required as the process is autocatalytic. An autocatalytic process is one in which the end products of the reaction themselves catalyse the reaction. In lipid autoxidation, the process starts with an initiation step in which a lipid molecule (R_1H) is dissociated to a lipid free radical (R_1*) and a hydrogen radical (H*). Once initiated, subsequent steps of oxidation follow automatically. The lipid free radical (R_1*) reacts with oxygen (O_2) to form the product, a peroxy free radical (R_1OO*) that, in turn, can produce another free radical (R_2*) by reacting with a second lipid molecule (R_2H). Thus autoxidation of lipids takes place through a 'free radical mechanism' as it starts and continues with the formation of free radicals. 'A free radical' is the component of a molecule, which has a free hand available due to removal of hydrogen radical from it. This free hand is free to bond with any reactive substance like oxygen. Lipid autoxidation is also a 'free radical chain reaction', because the product of the reaction of one molecule initiates the reaction of a second molecule, thus continuing the reaction in a chain.

Autoxidation of lipids takes place in four steps as follows.

1. Initiation
2. Propagation
3. Decomposition
4. Termination

1. Initiation

It is the first step in the autoxidation of lipids. In this step one hydrogen radical (H*) is removed from an unsaturated fatty acid of a lipid molecule (R_1H) at the carbon atom next to the double bond (*i.e.* the α-carbon to the double bond) resulting in a lipid free radical (R_1*) (Figure 56). This reaction is brought about by agents like light and metal ions. These agents, which initiate the formation of free radicals during the initiation step are called 'initiators' or 'chain initiators'. It is only their initiation after which the chain reaction can occur.

TEXT BOX 7
Lipid Deterioration

What is Lipid Deterioration ?

Lipid is an important component of fish tissues. It deteriorates after death of fish as well as during processing and storage of fishery products and byproducts. This is called 'lipid deterioration'. In general, lipids deterioration in fish and other foods is of four types as described below.

1. Hydrolytic rancidity (due to hydrolysis of lipids or lipolysis)
2. Oxidative rancidity (due to lipid oxidation or autoxidation)
3. Flavour reversion (change of flavour)
4. Polymerisation (formation of lipid polymers)

Out of the four types of lipid deterioration, oxidative rancidity is the most important from spoilage point of view of fish and fishery products. It is the rancid flavour resulting from the oxidation of lipids by atmospheric oxygen through the process called 'lipid oxidation' or 'autoxidation'.

1. Hydrolytic Rancidity

It is the rancid flavour resulting from the accumulation of free fatty acids (FFA) produced by hydrolysis of tissue fats by (a) lipolytic body enzymes (lipases) during autolysis and (b) lipolytic bacteria during microbial putrefaction (Figure 52). The free fatty acids produced by the hydrolysis of tissue lipids (fats) accumulate in fish tissues, which causes two problems as given below.

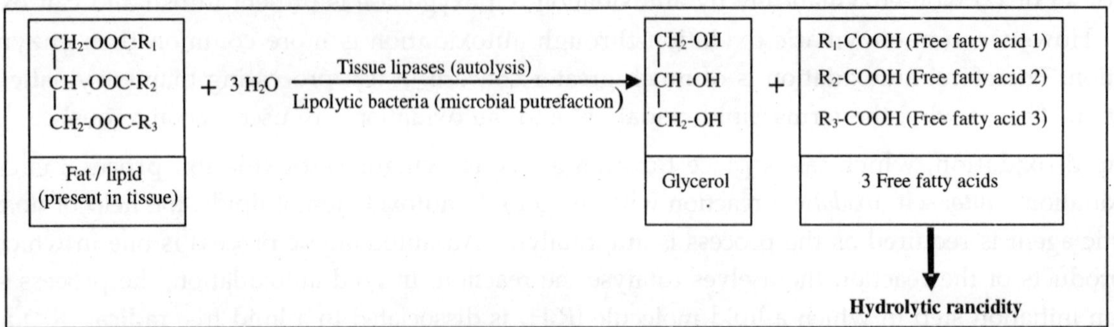

Figure 52: Hydrolytic rancidity

(a) Rancid Flavour

Presence of FFA imparts rancid flavour to the product.

(b) Creates Problems in Hydrogenation of Oil

FFA gets extracted into fish oils (both body oil and liver oil) during their industrial production. Its presence in fish oil is unwanted as it reduces the speed of hydrogenation by acting as 'catalyst poison' to the catalysts used to catalyse the hydrogenation process. Therefore, before hydrogenation, fish oil is subjected to the process of 'refining' for the removal of FFA. During the process, fish oil is agitated with an alkali. The FFA react with the alkali to form ' soap molecules' which separate out and settle to the bottom. The FFA- free oil is collected from the top (Figure 53).

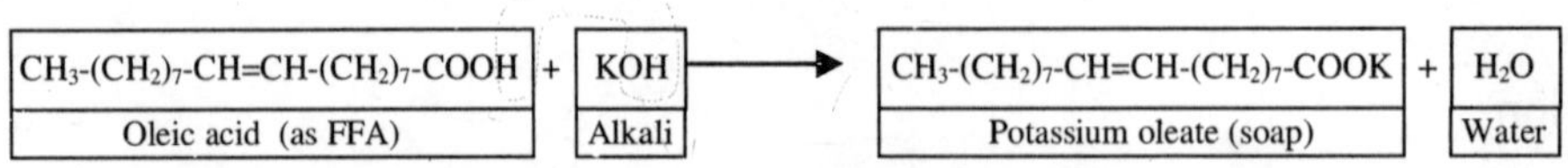

Figure 53: Refining process of oil (Example shown for oleic acid as the FFA)

2. Oxidative Rancidity

It is the rancid flavour, which develops in fish after death or in fishery products during processing and storage due to accumulation of certain end products of autoxidation (described in this chapter in details).

Contd...

Text Box 7–Contd...

3. Flavour Reversion

Lipids containing the fatty acid 'linolenic acid' (Figure 54) undergo a type of deterioration called 'flavour reversion' when exposed to air. Flavour reversion is the development of a particular type of oxidised off-flavour, which develops at comparatively low levels of lipid oxidation. This off-flavour is described as "greasy', "beany", "fishy" or "painty". It is produced with considerably less oxygen than autoxidation. The origin of this off-flavour appears to be the volatile oxidation products resulting from terminal pentane radical (CH_3-CH_2-CH=CH-CH_2-) of linolenic acid. A number of lipid oxidation end products, which are volatile compounds such as aldehydes, ketones, alcohols, esters and acids are present in oil, which has undergone flavour reversion. As fish oil contains linolenic acid, though at a small level of less than 2% of total lipids, it is susceptible to flavour reversion.

CH_3-CH_2-CH=CH-CH_2-CH=CH-CH_2-CH=CH-$(CH_2)_7$COOH

Terminal pentane radical (penta = 5 carbon)

Figure 54: Linolenic acid

4. Polymerisation

It is a type of lipid deterioration, which occurs only when lipids are exposed to high temperature processing such as canning and frying. In the process of polymerisation, single units of fatty acid monomers combine in a chain manner to form a chain or polymer of fatty acids. The unsaturated fatty acid of one fat molecule (say a triglyceride) forms bond with the unsaturated fatty acid of another fat molecule at the point of unsaturation of both the fatty acids (*i.e.* point where there are double bonds) (Figure 55). In canning of fish, the fish tissue lipids as well as the oil used as filling media in case of fish-in-oil cans may undergo polymerisation when subjected to high temperature during retorting. In frying of fish in vegetable oil, both the fish tissue lipids as swell as the frying oil undergo polymerisation when subjected to high temperature during the frying process. Here polymerisation results in an increase in the viscosity of the frying oil due to the increase in molecular size of the lipid molecules resulting from polymerisation. Increase in viscosity may greatly enhance foaming of the frying oil. This eventually limits the usefulness of the oil, as extensive foaming may make further frying in it impossible. However, fish absorbs some amount of oil during frying. This oil is replenished with fresh oil, which often reduces the effect of polymerisation.

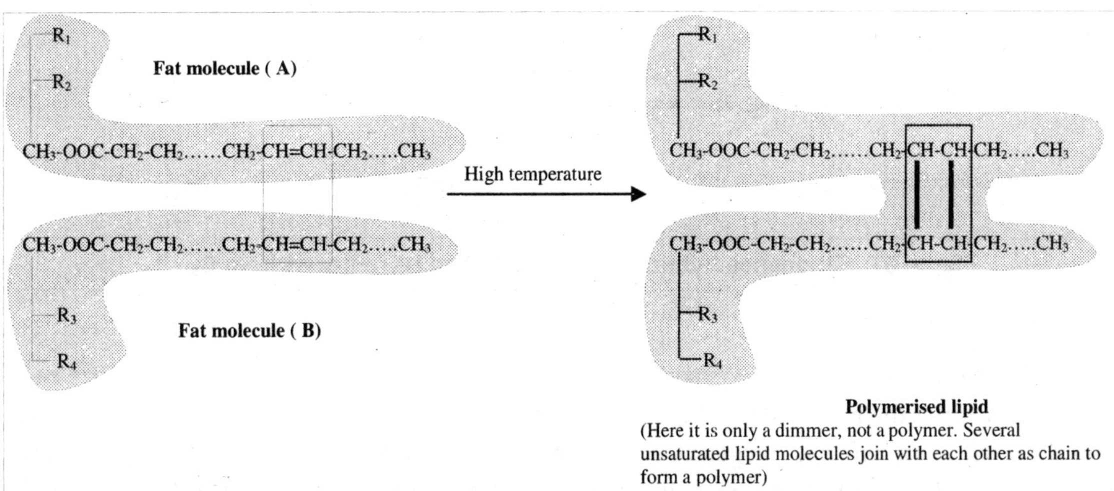

Figure 55: Polymerisation of lipids (A simplified representation)

The free radical formed in the initiation step is stabilised by 'resonance'. That means, the radical is stabilised by the movement of the free hand from one atom to another. For example, it moves from C_8 to C_{10} and from C_{10} to C_8 ($8 \leftrightarrow 10$) or from C_9 to C_{11} and C_{11} to C_9 ($9 \leftrightarrow 11$) of the fatty acid.

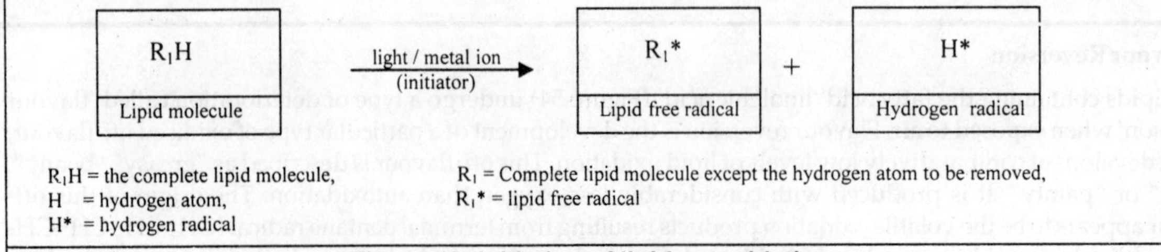

Figure 56: Initiation of lipid autoxidation

To elucidate resonance stabilisation of lipid free radicals, let us take the example of a fatty acid, oleic acid[1] in a fat molecule. Oleic acid has one double bond between C_9 and C_{10}. The initiation reaction proceeds by abstraction of H from C_8 or C_{11} as they are the carbon atoms next to the double bond. When H is removed from C_8, one pair of resonance hybrid (8 ↔ 10) results (Figure 57). In one of the pairs, the free hand is on C_8 and the double bond is on C_9-C_{10}, while in the other, the free hand shifts to C_{10} and the double bond shifts to C_8-C_9. Similarly, when H is removed from C_{11} another pair of resonance hybrid (9↔11) results (Figure 58). Similar event also occurs here. In one of the resonance hybrid pairs, the free hand is on C_{11} and the double bond on C_9-C_{10} while in the other the free hand shifts to C_9 and double bond shifts to C_{10}-C_{11}. Thus, for a fatty acid with only one double bond, two pairs (A&B and C&D) of resonance hybrids are produced (Figures 57 and 58).

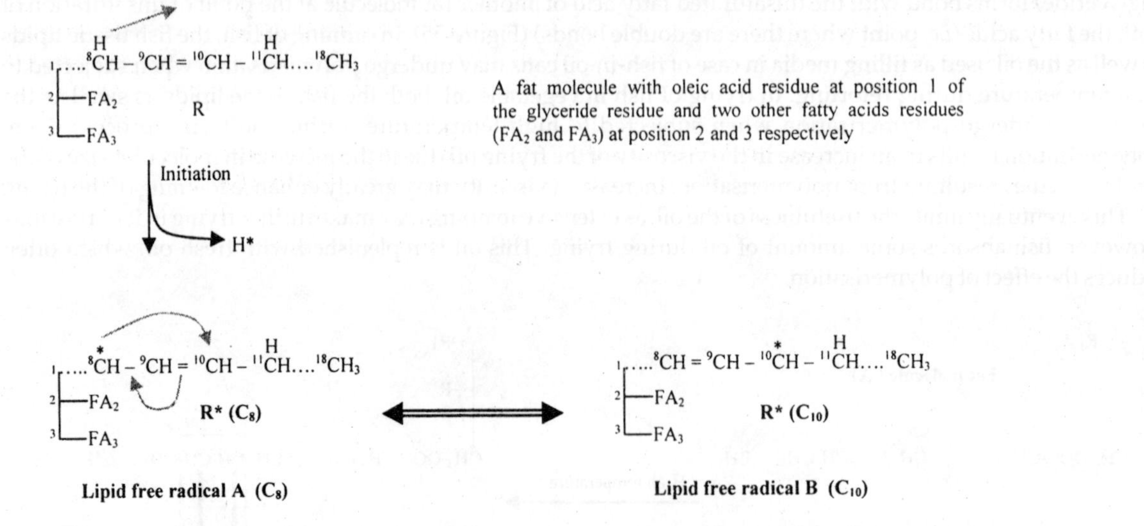

Figure 57: Resonance hybrid of oleic acid when H* removed from C_8

2. Propagation

In this step, the few free radicals formed in the initiation step propagate to produce large number of free radicals. At first, the free radicals (R_1*) formed in the initiation step from the first lipid molecule (R_1H) combines with atmospheric oxygen (O_2) to form the first peroxy free radical (R_1OO*) (Figure 59). Then this peroxy radical (R_1OO*) abstracts hydrogen radical (H*) from the unsaturated fatty acid of a second lipid molecule (R_2H) to produce a hydroperoxide (R_1OOH) and a new free radical (R_2*).

While the hydroperoxide (R_1OOH), which is not a free radical is stable; the new free radical (R_2*) is highly reactive and combines with oxygen (O_2) to form the second peroxy free radical (R_2OO*)

1 Oleic acid has been taken as the representative fatty acid to explain autoxidation here, because (1) it is present in the tissue of most fishes though not in large amount and (2) it has a single double bond for which it is be easy to explain the oxidation process. If other unsaturated fatty acids with more than one double bonds are taken, the process becomes complex and difficult to explain.

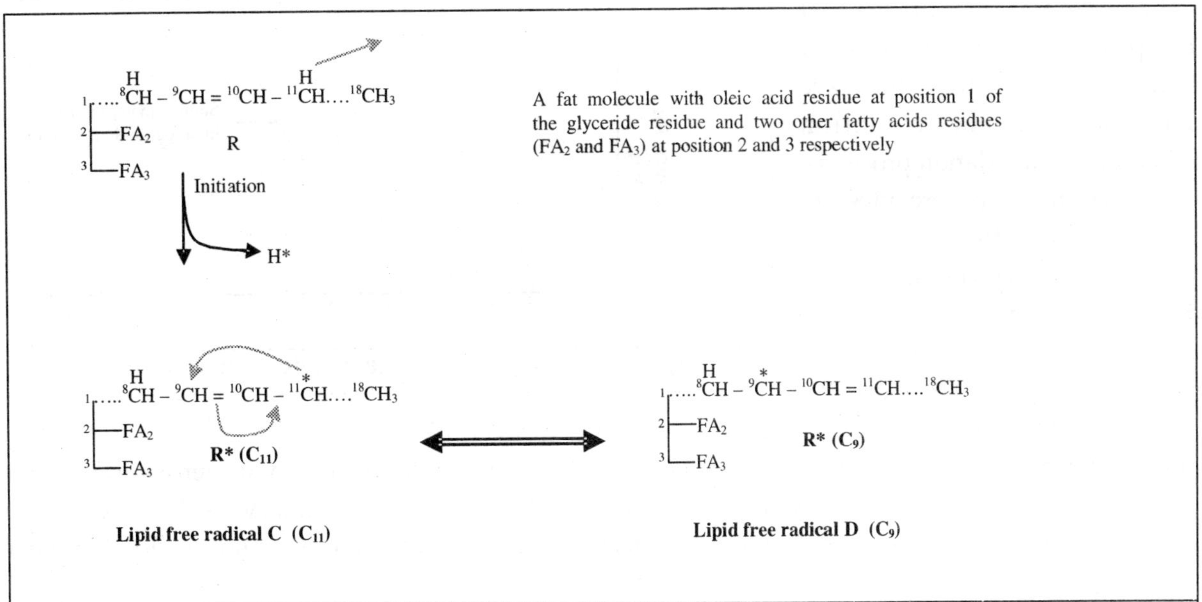

Figure 58: Resonance hybrid of oleic acid when H* removed from C_{11}

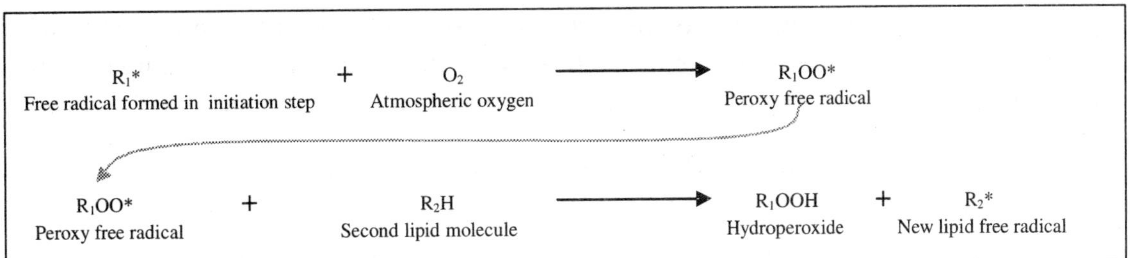

Figure 59: Propagation step in lipid autoxidation

(Figure 60). Then this peroxy radical (R_2OO^*) abstracts hydrogen radical (H*) from the unsaturated fatty acid of a third lipid molecule (R_3H) to produce a hydroperoxide (R_2OOH) and a new free radical (R_3^*). This reaction is repeated thousands of times and has the nature of a chain reaction (Figure 60).

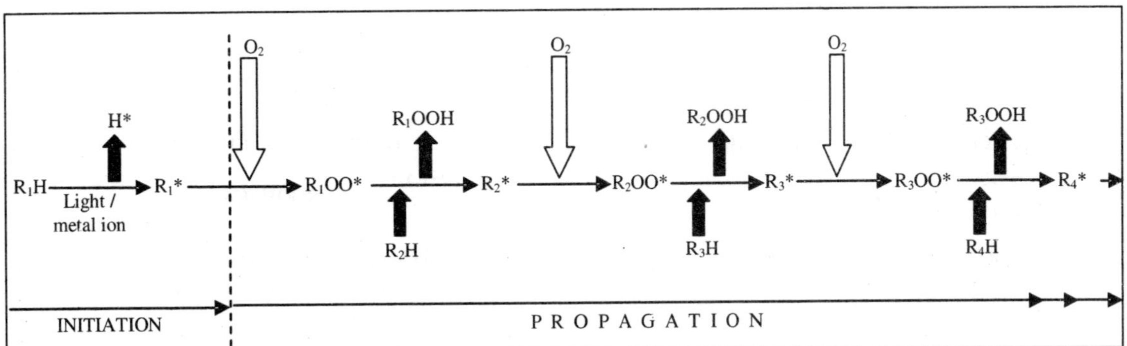

Figure 60: Chain reaction in the propagation step of lipid autoxidation (A very small portion shown)

Autoxidation, once initiated, will proceed not only unchecked, but also in a constantly accelerating manner brought about by the introduction of free radicals through peroxide break down and mediated by the availability of reactants. Therefore, in the initial period of the propagation step, the hydroperoxide formation takes place slowly. This period is called 'induction period'. After this period, the hydroperoxide formation increases suddenly (Figure 61).

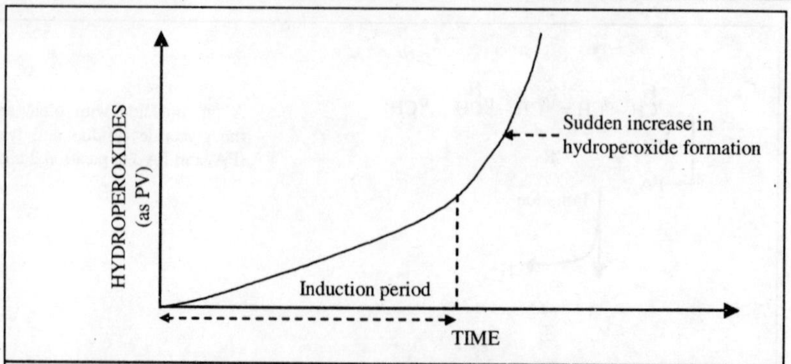

Figure 61: Change in the rate of autoxidation with time

The hydroperoxides are the first stable products in the process of autoxidation. Therefore, they are called the 'primary oxidation products'. As autoxidation proceeds, more and more hydroperoxides are produced. The hydroperoxides are a type of peroxide. The amount of hydroperoxides formed can be easily determined and expressed as 'peroxide value' (PV). Therefore, PV is used as an indicator of the extent of lipid oxidation. However, PV is not always an accurate measure of the extent of autoxidation, because in the advanced stage of autoxidation, the hydroperoxides undergo decomposition and correspondingly the PV decreases. Moreover, these primary oxidation products cannot impart the oxidative rancid flavour to a detectable extent and therefore are not so closely related to sensory properties.

In case of autoxidation of lipid containing oleic acid residue, the hydroperoxides produced in the propagation step has four isomers[2] derieved from the four sets of free radicals produced in the initiation step (Figure 62). Besides these four isomers, *cis-trans* isomerism also takes place and 90 per cent of the hydroperoxides formed are converted from their *'cis'* form to *'trans'* form isomers (Figure 63). Thus, for oleic acid with only one double bond, eight isomeric hydroperoxides are formed (4 *'cis'* isomers and 4 *'trans'* isomers).

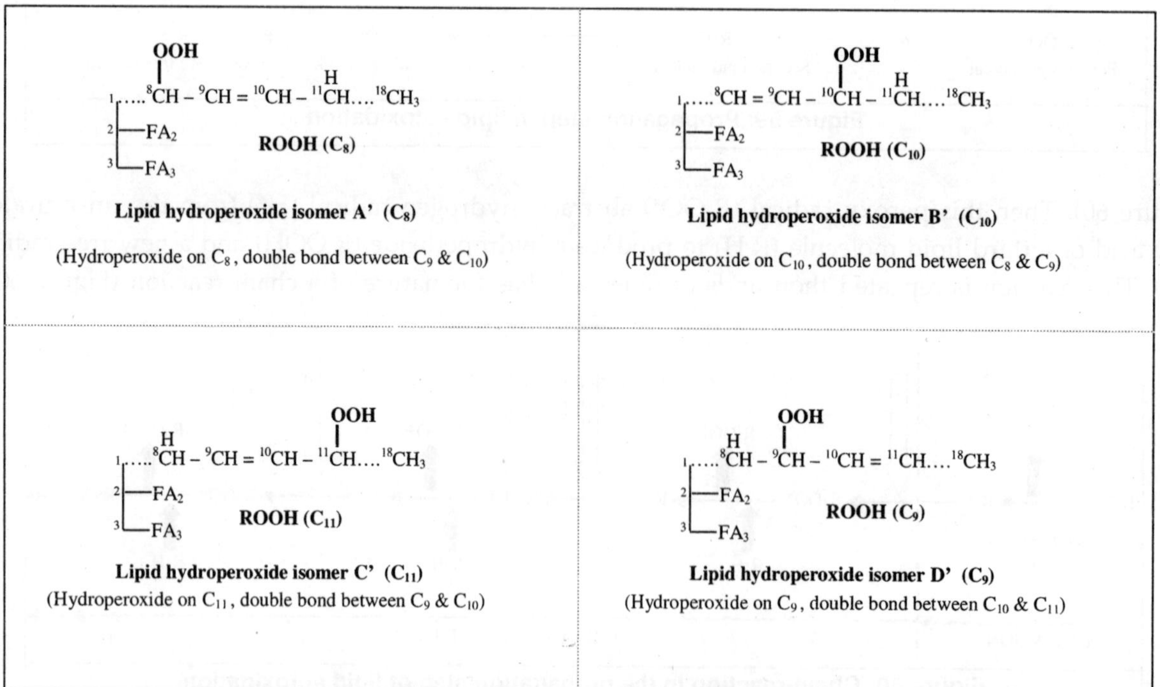

Figure 62: The four hydroperoxide isomers of lipid containing oleic acid at position 1 (Refer Figure 57 and 58)

Before the present free radical mechanism was clearly understood, it was thought that, oxygen molecule is directly added to the double bond in the fatty acid residue of the fat molecule to produce a type of peroxide as given in Figure 64. However, the present mechanism is based on the formation of another type of peroxide, a hydroperoxide, in which the original number of double bonds is retained.

---

2 Isomers are substances having same molecular formula, but different molecular structures.

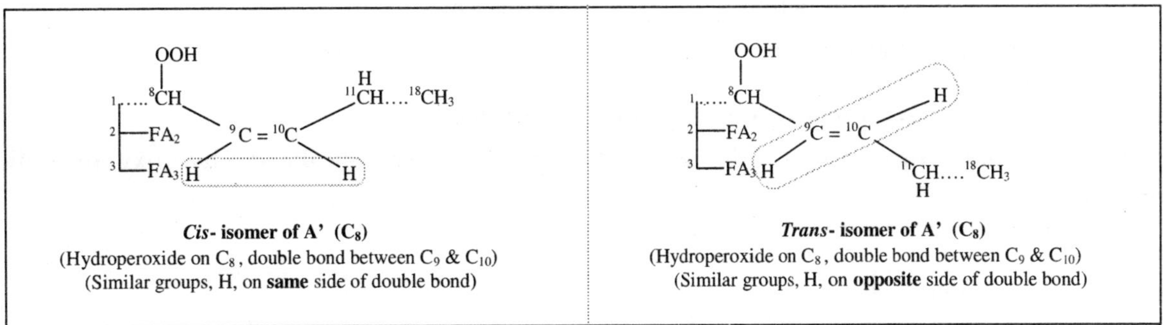

Figure 63 : *Cis –Trans* isomers of lipid containing oleic acid at position 1

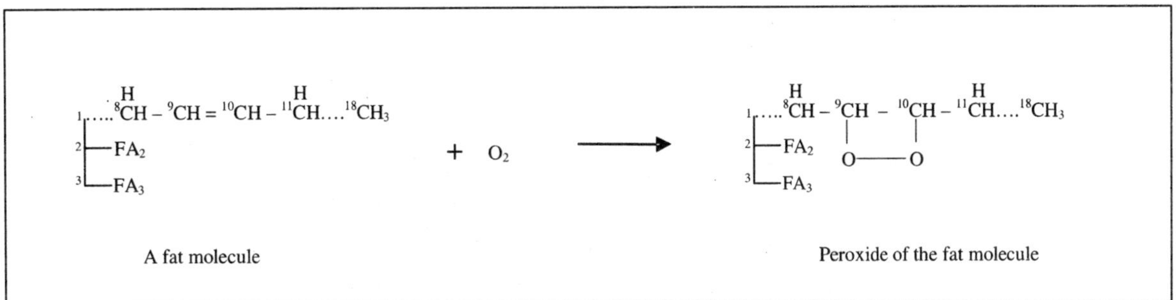

Figure 64: Formation of lipid peroxide (Old concept)

3. Decomposition

The hydroperoxides formed during propagation are the primary oxidation products. They are generally not very stable. In the decomposition step, these feebly stable hydroperoxides are decomposed into 'secondary oxidation products' (Figure 65). The secondary oxidation products such as aldehydes are sometimes decomposed further to produce 'tertiary oxidation products' such as short chain fatty acids. Both, the secondary and tertiary oxidation products are stable.

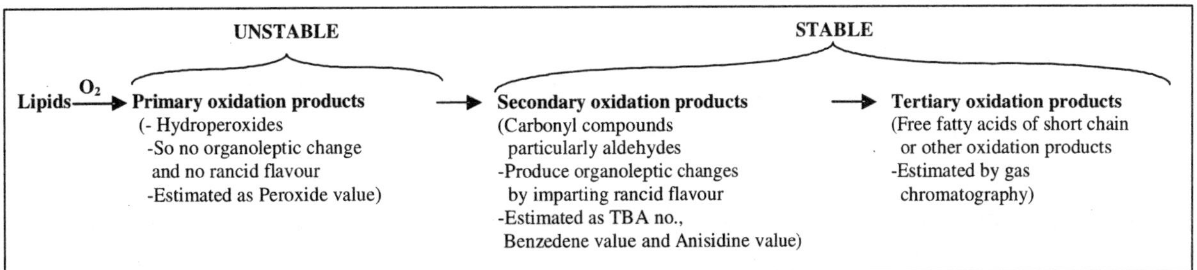

Figure 65: Products of lipid autoxidation

The secondary oxidation products include a variety of compounds with 'carbonyls' being the most important. The primary oxidation products do not develop rancid flavour to an objectionable level. It is the secondary oxidation products such as carbonyls, particularly the carbonyl, 'aldehydes', which produce the rancid flavour to a great extent. Therefore, these secondary oxidation products closely relate to sensory changes (organoleptic changes) *i.e.* the rancid flavour as sensed by man. One of the secondary oxidation products is 'malonaldehyde', which is an aldehyde. Its concentration in lipid indicates the extent of lipid autoxidation and is expressed as 'TBA number' (thiobarbituric acid number). Malonaldehyde reacts with 2-thiobarbituric acid to produce red colour which is estimated by a spectrophotometer. Besides TBA number, 'benzedene value' and 'anisidine value' are also estimated to determine the secondary oxidation products. The hydroperoxides formed in the propagation step are decomposed in two steps as follows.

A. Decomposition of hydroperoxide to alkoxy free radical

B. Decomposition of alkoxy free radical to aldehyde, alcohol or ketone

A. Decomposition of Hydroperoxide to Alkoxy Free Radical

In this step, lipid hydroperoxide is decomposed to an alkoxy free radical and a hydroxy free radical (Figure 66).

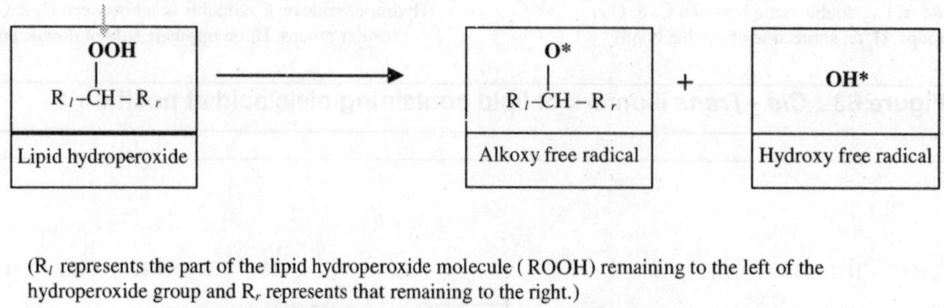

Figure 66: Decomposition of lipid hydroperoxide

For example, in case of oleate (lipid molecule containing oleic acid), the first step of decomposition takes place breaking the oleate hydroperoxide to alkoxy free radical of oleate and hydroxy free radical (Figure 67).

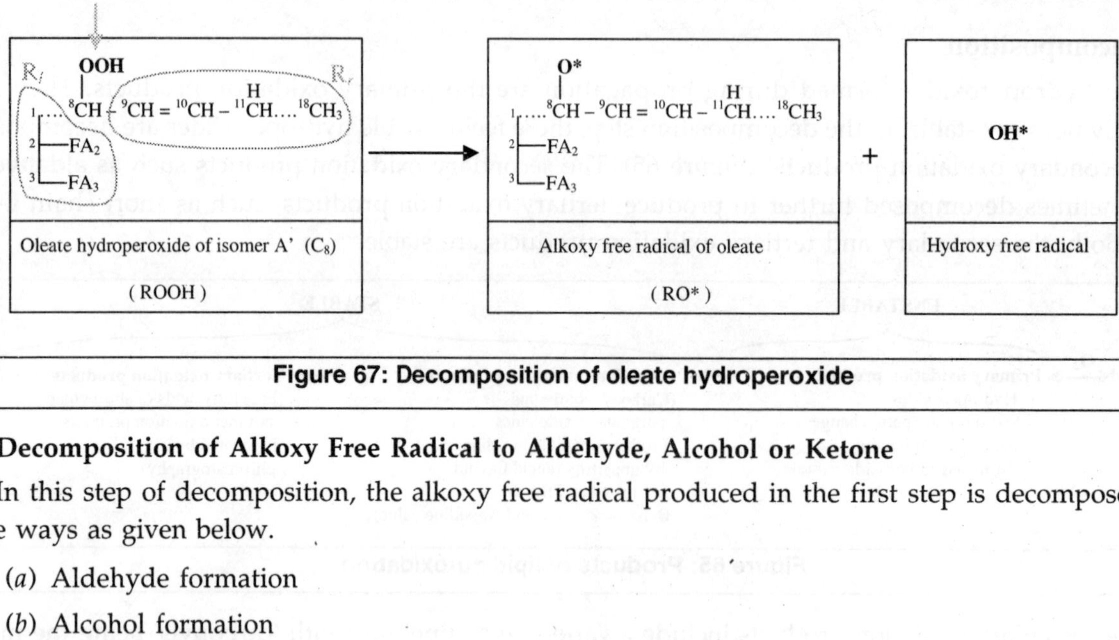

Figure 67: Decomposition of oleate hydroperoxide

B. Decomposition of Alkoxy Free Radical to Aldehyde, Alcohol or Ketone

In this step of decomposition, the alkoxy free radical produced in the first step is decomposed in three ways as given below.

(*a*) Aldehyde formation

(*b*) Alcohol formation

(*c*) Ketone formation

(a) Aldehyde Formation

The alkoxy free radical may decompose to produce an aldehyde and a lipid free radical. The formation of the aldehyde from the alkoxy free radical involves the break down of the chain of fatty acid on the fat molecule. It may occur either on the left side or right side of the free radical. In this way two types of aldehydes may be produced (Figure 68).

For example, in case of oleate, chain breaking may occur at 7-8 bond or 8-9 bond, *i.e.* at position A or B (Figure 69). Accordingly, if the fatty acid chain breaks at A, the aldehyde formed (R_r CHO) is a short chain volatile aldehyde and if it breaks at B, the aldehyde formed (R_l CHO) is a non-volatile aldehydoglyceride. The volatile aldehydes, which are secondary oxidation products, are mainly

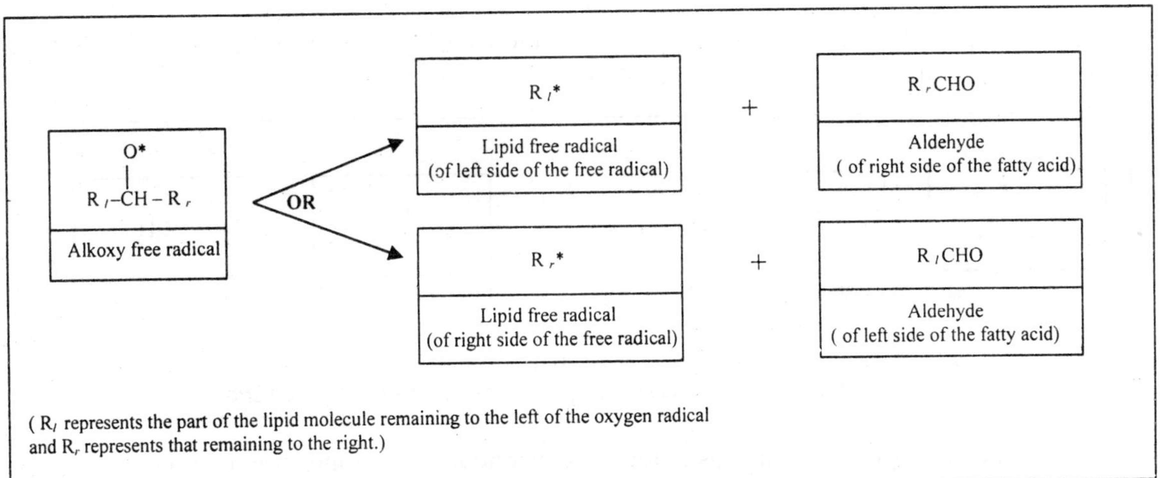

Figure 68: Decomposition of alkoxy free radical to aldehyde

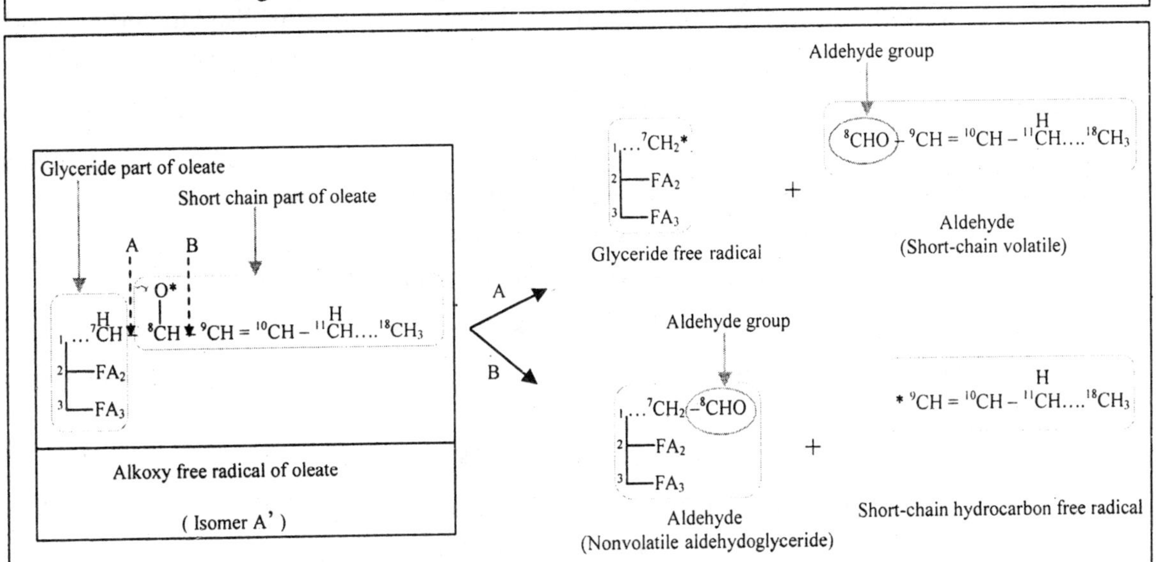

Figure 69: Decomposition of alkoxy free radical of oleate to two different aldehydes

responsible for the rancid flavour of oxidised lipids. However, the volatile oxidation products, such as the short chain volatile aldehydes, can be removed in the refining process during deodourisation of fish oils, but the non-volatile products such as the aldehydoglyceride, remain in the oil and this results in a lower oxidative stability of fish oils.

(b) Alcohol Formation

The alkoxy radical may also abstract a hydrogen from another lipid molecule (R_2H) to yield an alcohol and a new free radical (R_2*). This new free radical may participate in propagation of the chain reaction or interact with another alkoxy free radical to terminate the chain (Figure 70).

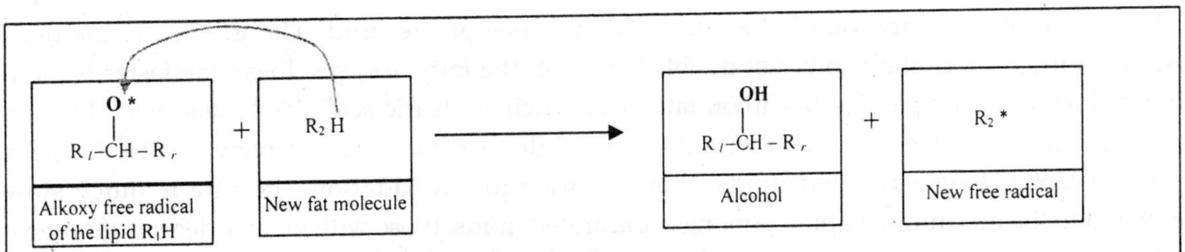

Figure 70: Decomposition of alkoxy free radical to alcohol

(c) Ketone Formation

The alkoxy free radical may react with a new free radical (R_2*) formed in the above step to terminate the chain reaction with formation of ketone (Figure 71).

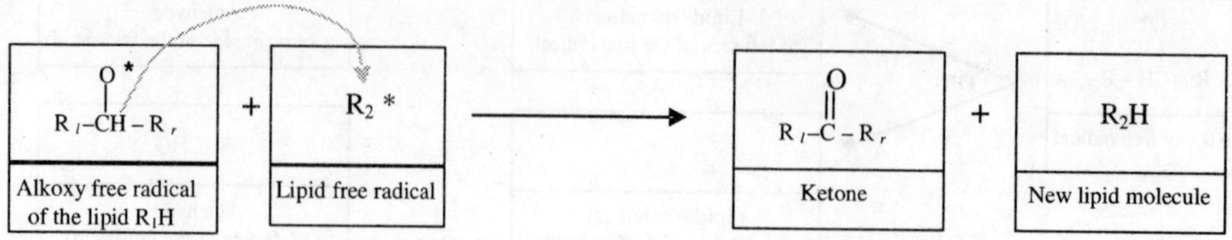

Figure 71: Decomposition of alkoxy free radical to ketone

The decomposition products such as aldehydes, alcohols and ketones impart rancid flavour to the oxidised lipids. Among them, the aldehydes impart the highest rancid flavour. If the aldehyde has double bond, the rancid flavour is still more. The aldehydes can be further oxidised to 'tertiary oxidation products' such as carboxylic acids like free fatty acids or other tertiary oxidation products (Figure 65).

4. Termination

The chain reaction of autoxidation does not continue indefinitely. It is stopped or terminated in the termination step, which is the last step of the autoxidation process. In this step, the free radicals react within themselves to yield non-reactive products as follows.

A. A pair of free radicals may react to form a stable compound.

$$R_1^* + R_2^* \longrightarrow R_1 - R_2$$

B. A free radical may react with a peroxy radical produced in the first phase of propagation step to form a stable compound.

$$R_1^* + R_2OO^* \longrightarrow R_1OOR_2$$

C. A pair of peroxy free radicals may react to form a stable compound.

$$R_1OO^* + R_2OO^* \longrightarrow R_1OOR_2OO$$

D. A large number (n) of peroxy free radicals may react to form a stable product.

$$n\ (ROO^*) \longrightarrow (ROO)_n$$

III. Factors Influencing Lipid Autoxidation

Autoxidation of lipids is influenced by the following factors.

1. Composition of the Lipids

Composition of tissue lipids of fish with respect to the following aspects has great influence on the rate of lipid autoxidation.

(a) Degree of Unsaturation

The most important factor which influences the rate of lipid oxidation and rancidity development is the degree of unsaturation of the fatty acid residues of the lipid. The greater is the degree of unsaturation (*i.e.* more the number of double bonds on the fatty acid residues), the faster is the rate of autoxidation. For example, the 18 carbon fatty acids such as stearic acid (18:0), oleic acid (18:1), linoleic acid (18:2) and linolenic acid (18:3) with 0,1,2 and 3 double bonds respectively oxidise in the rate of 1:100:1200:2500. Thus saturated lipids, though undergo autoxidation, the rate is much slower as compared to the unsaturated lipids. Among unsaturated lipids, those with greater degree of unsaturation oxidise at much faster rate than those with lesser degree of unsaturation.

Tissue lipids of fish, particularly those of the marine organisms (often called 'marine oils') are rich in lipids containing fatty acids with high degree of unsaturation. They are called polyunsaturated fatty acids (PUFA) containing more than one double bond and highly unsaturated fatty acids (HUFA), containing large number of double bonds. That is why, fish tissue lipids oxides very quickly and cause rancidity deterioration of fish.

Though more unsaturated lipids oxidise very fast, phospholipids, which are the most unsaturated lipids in all fishes, do not oxidise rapidly because of their physical disposition. They are mostly present in the cell membrane as its integral component. As they are bound in the membrane matrix, they have less mobility, which makes it difficult for them to move from molecule to molecule and participate in the propagative oxidation chain reaction.

(*b*) Geometry of Double Bonds

The geometry of the double bonds (*i.e.* the spatial configuration- configuration in space-, of the groups or atoms attached to the double bonds) results in *cis*- and *trans*- isomers of the fatty acids (Figure 62). The lipids with *cis* fatty acid isomers oxidise faster than those with *trans* fatty acid isomers.

(*c*) Position of Double Bonds

Lipids having conjugated (*i.e.* very close) double bonds oxidise faster than those with non-conjugated ones in which the double bonds are away from each other.

(*d*) Disposition of the Fatty Acids within the Lipid Molecule

In case of fish lipids, very often, a triglyceride or fat molecule contains three different types of fatty acids bonded to the glycerol molecule. Among them, the two outer fatty acid residues bonded to the glycerol at positions 1 and 3 oxidise more rapidly than the one attached at position 2. The two outer fatty acids protect the inner fatty acid by exposing themselves to the attack by atmospheric oxygen.

(*e*) Freedom of Fatty Acids

Fatty acids oxidise at a slightly greater rate when free than when esterified to glycerol. That mean, the free fatty acids (FFA) produced by the hydrolysis of fats are more readily oxidised.

2. Temperature

Temperature has marked influence on the rate of lipid autoxidation. The rate of autoxidation increases with increasing temperature, as rise in temperature activates reacting molecules and at the same time promotes the decomposition of the hydroperoxides. The rate of autoxidation also decreases with decreasing temperature usually by a factor of 2 to 3 for every 10°C fall in temperature. However, unlike microbial activity, which is completely inhibited below a temperature of –10°C, autoxidation is not completely inhibited even at very low temperatures of freezing and cold storage of fish. At the usual freezing temperatures of –35°C to –40°C and at the normal cold storage temperature of less than –18°C, autoxidation of fish tissue lipids continues unabated though at a very slow rate.

In fish lipids, autoxidation proceeds quite readily at ambient temperature. However, in case of chilled fish and fresh fish to be marketed fresh at ambient temperature (*i.e.* wet fish), the deterioration of quality of fish is caused by microbial and enzymatic activities to such an extent that, the deterioration caused by autoxidation and resulting rancidity development becomes insignificant. Fish spoils before the development of any detectable rancidity. But in case of frozen fish, during cold storage, microbial and enzymatic activities are almost completely inhibited whereas autoxidation continues though at a much slower rate. Therefore, oxidative deterioration is the most important agent of quality loss in frozen fish during cold storage. Autoxidation, which dominates the quality deterioration of fish during cold storage, is further enhanced by long-term ice storage (chilling) before fish is frozen as well as by pre-freezing brining and cold storage dehydration.

3. Degree of Exposure of Flesh

Autoxidation takes place at a faster rate if the tissue lipids are more exposed to atmosphere. As oxygen diffuses very slowly through the fish tissue, the lipids, which are embedded deep in the tissue unexposed to atmospheric oxygen get oxidised very slowly. That is why, when more surface area of fish tissue is exposed to air by skinning, filleting, mincing, gutting or preparing fish meal, the tissue lipids get oxidised quickly. Similarly, in frozen fish, to avoid direct exposure of fish flesh to atmospheric oxygen during cold storage, certain methods are followed. (1) In case of block-frozen shrimp or fish, before freezing, they are given a 'pre-freezing glaze treatment' in which fish or shrimp are arranged in a container (usually a rectangular tray) and then chilled water is poured into the container. This water fills the void space between shrimps or fishes and completely covers the product from all sides thereby giving a uniform block after freezing. This water is called 'glaze water'. As it covers the product from all sides, it does not allow direct contact of the product with atmospheric oxygen and thereby prevents autoxidation. (2) In case of individual frozen products such as individually quick frozen (IQF) shrimps or larger fishes frozen by air blast freezing, after freezing is over, the frozen product is given a 'post-freezing glaze treatment' by spraying chilled water to its surface or by giving it a quick dip treatment in chilled water. Due to the very low surface temperature of the product after freezing, a thin layer of water quickly solidifies on its surface and forms a coating on it. This prevents contact with atmospheric oxygen and thus prevents autoxidation. It can also be adopted in case of block frozen products. (3) Frozen fish is packaged properly using packaging materials having good gas barrier properties, which prevents contact of the product with atmospheric oxygen. Besides preventing lipid autoxidation, all the three steps also prevent 'dehydration'. Dehydration is the loss of water from frozen products during cold storage, which lead to quality deterioration of the products.

In fishmeal, exposure of large surface area as well as high temperature treatment result in so rapid autoxidation of its lipids that it may sometimes burn. Also, in fish meal prepared from fatty fishes little unreacted PUFA is left over. However, in fishmeal, other oxygen consuming reactions account for upto 10 times the oxygen consumed by autoxidation.

4. Oxygen Concentration

If oxygen supply is unlimited, the rate of autoxidation does not depend upon oxygen concentration. But when oxygen concentration is very low, the rate of autoxidation is proportional to the concentration of oxygen. For example, in case of fish and other foods preserved by canning, as air is completely removed from inside the tin can before it is closed, such products never undergo autoxidation. For the same reason, many other fishery products are packed in oxygen-free atmosphere. In such case, they are packed with nitrogen gas, which excludes air and does not allow lipids to get oxidised. Packing fish in impermeable bags filled with N_2, CO_2 or deoxygenated air strongly inhibits lipid oxidation. This method is adopted in the modified atmosphere packaging (MAP) of fishes. In case of lean fishes all the three gases CO_2: N_2:O_2 are used in the ratio of 40 per cent: 30 per cent: 30 per cent respectively. But in case of fatty fishes, O_2 is not used as it oxidises the tissue fats and turn them rancid. Here, oxygen is excluded from the pack by flushing into the pack, CO_2 and N_2 in the ratio of 40-60 per cent and 60-40 per cent respectively.

5. Fat Content of Fish, its Types and Disposition

Fatty fishes contain large amount of lipids[3] consisting mainly of triglycerides (fats) deposited in the 'adipocytes' or 'fat cells' of the 'adipose tissue' and also within the muscle cells. This deposit of fats is called 'depot fats'. The triglyceride molecules of the depot fats remain free inside the cells and therefore have greater mobility. That is why, once autoxidation of lipids starts, the greater mobility of the

3 Lipids are mainly of two types:

1. Fats or triglycerides present in the fat depots.
2. Phospholipids present in the cell membrane of all cells.

triglyceride molecules inside the cells helps in rapid propagation of autoxidation process and results in fast rate of autoxidation. On the contrary, lean fishes contain small amount of lipids consisting mainly of phospholipids present as integral component of the cell membrane of all cells and are bound to the proteins in the cell membrane matrix. Therefore, they cannot move freely. This restricted mobility does not allow phospholipids to participate in the propagation of autoxidation and results in their slow autoxidation process, although they are the most unsaturated lipids. For this reason, fatty fishes oxidise faster than lean fishes; triglycerides oxidise faster than phospholipids and depot fats oxidise faster than membrane bound lipids. Fatty fishes like herring and mackerel, which possess large lipid reserves are most susceptible to autoxidation and this becomes the dominant agent of quality deterioration in frozen fish during cold storage. In contrast, fishes such as cod and lizard fish, which are deficient in such lipids, contain structural lipids (phospholipids) associated with the membrane. Therefore, in these fishes autoxidation is not a dominant deterioration process during frozen storage. Thus, the shelf life of frozen fatty fishes is less than that of frozen lean fishes.

6. Location of Lipids

Lipids of dark lateral muscles[4] of fish tend to oxidise faster than those of the white muscle. This is due the higher content of fat, heme pigments and TMA in the dark muscle, which enhance autoxidation.

7. Extraction

Lipids, when present in fish carcass, undergo rapid autoxidation than when extracted as fish oil. This is because; certain muscle constituents participate in the autoxidation process and enhance it. They are absent in the extracted fish oil.

8. Presence of Prooxidants

Certain substances are known to accelerate the rate of autoxidation. They are called 'prooxidants'. The presence or addition of prooxidants to fish leads to rapid autoxidation. Some of the important prooxidants are given below.

(*a*) Light and Ionising Radiations

They are powerful accelerators of lipid autoxidation. Light accelerates autoxidation by helping in the 'initiation step' of autoxidation resulting in the formation of hydroperoxides. It also accelerates the decomposition of the hydroperoxides. Therefore, when lipid materials are packaged in transparent containers, they undergo rapid autoxidation. Irradiation by ionising radiation enhances autoxidation because the microbes, which compete with lipids for oxygen, are destroyed by irradiation and also irradiation produces free radicals.

(*b*) Metals

Metals, especially copper, iron and cobalt catalyse autoxidation of lipids. Metal deactivators such as citric acid can be used to reduce this effect.

(*c*) Metallic Soaps

Metallic soaps are formed by the reaction of metallic components of metal containers in which fish oil is stored and the free fatty acids present in the extracted oil in small quantities. These soaps accelerate autoxidation of lipids.

(*d*) Common Salt

The common salt, sodium chloride (NaCl) has high prooxidant properties. Therefore, the practice of dipping fish fillets in salt solution (brining) before freezing in order to reduce drip loss[5], results in

4 Dark lateral muscle are the two dark bands of muscle present on both the sides of fish along the lateral lines.

5 Drip loss: Just before use, when solidified frozen fish is melted or defrosted by the process called thawing, some fluid is lost from the fish body. The loss of weight due to this drip of fluid from the fish tissue during thawing is called drip loss.

early rancidity of the fillets. Similarly, in brine freezing of fish in which fish is frozen by bringing it in direct contact with brine maintained at very low temperature (about–18°C), the flesh lipids come in direct contact with salt and get oxidised quickly. Also in preservation of fish by salting, the salt used accelerates the rate of autoxidation in the salted fish during storage. That is why; the above methods do not produce good quality products. To overcome this problem, antioxidants, which prevent autoxidation of lipids, are used alongwith the salt.

(*e*) Salt Impurities

The impurities present in salt are chlorides of magnesium, calcium, aluminium, zinc, beryllium, copper and iron. These salt impurities accelerate autoxidation of lipids and deteriorate the quality of salted fish.

(*f*) Moisture

Moisture has marked prooxidant effect when its level in fish is high (a_w^6 = 0.5-0.8), because high moisture content increases the mobility of the catalysts. On the other hand, lower levels of moisture (a_w = 0.1-0.5) protect lipids against autoxidation, as lower levels of moisture reduce the catalytic effect of metal catalysts by quenching free radicals or by impending the access of oxygen to fish tissue. Again, at very low levels of moisture ($a_w < 0.1$), autoxidation proceeds rapidly. The prooxidant nature of moisture is of great concern.

(*g*) Heme Compounds

Heme compounds (hematic compounds/heme proteins) act as catalysts of lipid autoxidation. Among them, the heme pigments such as haemoglobin, myoglobin, cytochrome c and allied compounds enhance autoxidation. During the autoxidation of lipids in fish tissue, the hematic compounds are chemically changed and their concentration decreases.

(*h*) Flesh Components

Some of the flesh components act as prooxidants. That is why, lipids present in fish oxidise faster than lipids extracted as fish oil.

(*i*) Lipoxygenase

The enzyme lipoxygenase (previously called lipoxidase) enhances the rate of lipid oxidation by oxidising the lipids through enzymatic oxidation instead of the non-enzymatic autoxidation path. However, non-enzymatic autoxidation of lipids is the major path of lipid oxidation while the enzymatic path is of meager importance.

9. Presence of Antioxidants

Many foods contain natural antioxidants (Text box 8). Among them tocopherols[7] (vitamin E) are the most important. They are present in large quantity in vegetable oils than in animal fats for which vegetable oils are more stable (do not develop rancidity) than animal fats. When oil is extracted from the fatty tissues, say muscle or liver of fish, some amount of natural antioxidants present in these tissues gets extracted into the extracted crude oil. But when the crude oil is subjected to the refining process, the natural antioxidants are lost. Hence, such refined oils are more susceptible to autoxidation (more unstable) than the crude oils. The deficiency of such refined oils is rectified by addition of commercial antioxidants. Stability of fish lipids is largely proportional to the tocopherol content which ranges from 40 mg / kg for sardine oil to 630 mg / kg for sable fish oil. Squalene present in fish lipids is also a natural antioxidant.

6 a_w: Water activity (a_w) is the ratio of vapour pressure of fish tissue to that of pure water. It is used to indicate the level of moisture available in fish tissue for microbial and chemical activities.

7 Tocopherols are of different types such as α-tocopherol, γ-tocopherol and δ-tocopherol.

TEXT BOX 8
Antioxidants

Antioxidants are substances, which can act against the autoxidation of lipids. They are very effective in slowing down the autoxidation process and in increasing the length of the induction period. Antioxidants are of three types as given below.

A. Natural Antioxidants

They are naturally present in many organisms. Some of them include tocopherol (vitamin E), squalene and lecithin. The most important among them is tocopherol. It is the presence of tocopherol in tissue, which prevents autoxidation of tissue lipids in living organisms though oxygen is circulated constantly throughout the body via blood circulatory system. Tocopherol is a phenolic substance.

B. Induced Antioxidants

They are formed when exposed to such processing as smoking and roasting. They are formed in the smoke produced by burning of wood. They are of less importance from application point of view.

C. Synthetic Antioxidants

They are synthesised by chemical reactions. Most of the commercial antioxidants are synthetic antioxidants. Most of the antioxidants contain an aromatic phenolic ring. Of the several compounds, which can prevent autoxidation, the two groups namely phenolic compounds and aromatic amines are the most effective. The aromatic amines are generally considered too toxic for food uses. Therefore, only phenolic compounds are used as antioxidants for food purposes. Those phenolic compounds, which can easily produce quinone, are active as antioxidants.

Some common natural and synthetic antioxidants are given below (Figure 72).

Natural Antioxidants	*Synthetic Antioxidants*	
Tocopherol (most common) Squalene Lecithin Resin guaiac	Butylated hydroxy anisole (BHA) Butylated hydroxy toluene (BHT) Nordihydroguaiaretic acid (NDGA) Propyl gallate (PG) 1,4 Benzoquinone	Dibutyl paracresol Isoamyl gallate 6-Ethoxy,2,2,4-trimethyl-1,2-dihydroquinoline (EMQ) Tert-butyl hydroxy quinone (TBHQ)

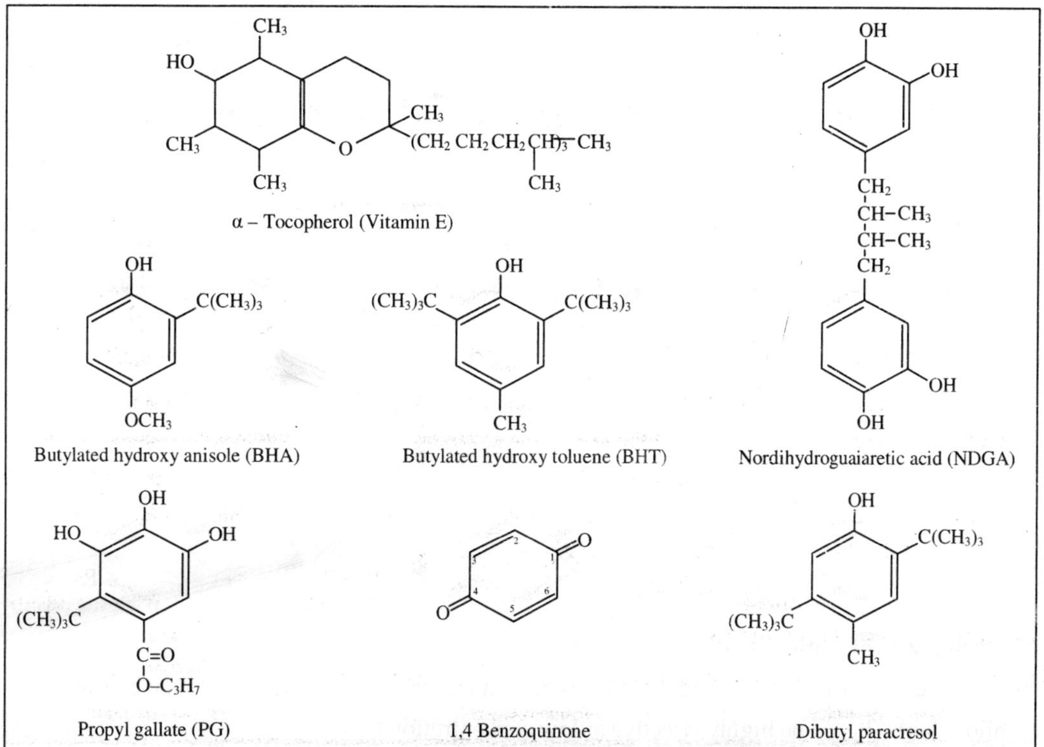

Figure 72: Molecular structure of important antioxidants

Contd...

Text Box 8–Contd...

The synthetic antioxidants are of utmost importance from application point of view. Most of them are phenolic compounds (i.e. contain phenolic ring). The number of antioxidants used in foods is limited. Some of them include BHA, BHT, PG, NDGA, TBHQ, EMQ, isoamyl gallate, resin guaiac and dibutyl paracresol. Among them, the most commonly used are BHA, BHT, PG and NDGA.PG is more soluble in water than in lipids. BHA has 'carry through' properties, but BHT does not have because it is volatile with steam.

Many foods contain natural antioxidants. Among them tocopherols (vitamin E) are the most important. Tocopherols are present in large quantity in vegetable oils than in animal fats for which vegetable oils are more stable (do not develop rancidity) than animal fat. When oil is extracted from the fatty tissues, say muscle or liver of fish, some amount of natural antioxidants present in the tissue gets extracted into the extracted crude oil. But when the crude oil is subjected to the refining process, the natural antioxidants are lost. Hence, such refined oils are more susceptible to autoxidation (more unstable) than the crude oils. The deficiency of such refined oil is rectified by addition of commercial antioxidants. Stability of fish lipids is largely proportional to the tocopherol content which ranges from 40 mg / kg for sardine oil to 630 mg/kg for sable fish oil. Squalene present in fish lipids is also a natural antioxidant.

Mechanisms of Antioxidant Action

There are four mechanisms by which antioxidants prevent autoxidation of lipids. They are as follows.

(*a*) Hydrogen donation by the antioxidants.

(*b*) Electron donation by the antioxidants.

(*c*) Addition of lipid to the aromatic ring of the antioxidants.

(*d*) Formation of complex between the lipid and the aromatic ring of the antioxidants.

Antioxidants prevent autoxidation by the removal of free radicals formed which leads to the termination of the chain reaction of autoxidation. There are three theories to explain how antioxidants remove free radicals formed during autoxidation as given below.

a) The antioxidant removes the free radical to its own molecule

R* (Lipid free radical) + AH (Antioxidant) → RH (Stable lipid) + A* (Antioxidant free radical)

OR

ROO* (Lipid peroxy free radical) + AH (Antioxidant) → ROOH (Stable lipid) + A* (Antioxidant free radical)

In another way,

ROO* (Lipid peroxy free radical) + AH_2 (Antioxidant) → ROOH (Stable lipid) + AH* (Antioxidant free radical)

AH* + AH* → AH_2 + A (Antioxidant)

OR

AH* + ROO* → ROOH + A (Stable lipid)

b) A complex is formed between the peroxy free radical (ROO*) and the antioxidant (AH). The free radical-antioxidant complex reacts with another molecule of peroxy free radical (ROO*) to yield stable lipid.

ROO* (Lipid peroxy free radical) + AH (Antioxidant) → ROO*- AH (Free radical- antioxidant complex) → ROOROO + AH (Stable lipid)

c) Here the antioxidant competes with the lipid molecules for free radical and thus does not allow further formation of lipid free radicals.

ROO* (Lipid peroxy free radical) + AH (Antioxidant) → ROOH (Stable lipid) + A* (Antioxidant free radical)

ROO* (Lipid peroxy free radical) + RH (Lipid molecule) ✗→ ROOH (Stable lipid) + R* (Lipid free radical)

Ideal Characteristics of Antioxidants

The ideal characteristics of antioxidants for use in fish and fishery products are given below.

1. Antioxidants should be highly effective at low concentrations.

2. It should have reasonable price.

Contd...

Text Box 8–Contd...

3. It should be easily soluble in lipid substances.
4. It should not produce any colour or flavour changes.
5. It should have no toxicity.
6. It should have effect on a wide range of lipid varieties.

Technique of Using Antioxidants

To prevent autoxidation of tissue lipids during frozen storage of fish, synthetic commercial antioxidants are added to it. It is important to note that, at high concentration, antioxidants may have a prooxidant effect. Therefore, addition of antioxidants beyond optimum amount may result in enhancement of lipid autoxidation.

In freezing of fish, antioxidants are usually added to the water used for pre-freezing glazing. They are often incorporated into the paraffin wax used for coating the cardboard cartoons in which frozen fish is packaged. In salted fish, antioxidants are added to the salt used for salting of fish.

To prevent autoxidation of tissue lipids during frozen storage of fish, synthetic commercial antioxidants are added to it. It is important to note that, at high concentration, antioxidants may have a prooxidant effect. Therefore, addition of antioxidants beyond optimum amount may result in enhancement of lipid autoxidation.

10. Presence of Synergists

The presence of synergists in lipid tissues or extracted oils enhances the prevention of autoxidation by the added antioxidants (Text box 9).

11. Cold Storage Dehydration

Dehydration of frozen fish during cold storage is the loss of water from fish body to the cold store atmosphere when the relative humidity inside the cold store is less than that of the frozen fish. Loss of water through dehydration results in direct exposure of fish tissue to atmospheric oxygen, which accelerates autoxidation. Thus, dehydration accelerates autoxidation. It can be overcome by preventing dehydration through freezing fish after filling the void space with water (pre-freezing glazing), by creating a layer of ice on the frozen fish (post-freezing glazing) and by proper packaging of frozen fish.

TEXT BOX 9
Synergists

Synergists are substances, which are not highly effective by themselves as antioxidants, but when used together with antioxidants, increase their effectiveness in preventing lipid autoxidation. Some of the important synergists are phosphoric acid, citric acid, tartaric acid, isopropyl citrate, ascorbic acid, ascorbyl palmitate and various organic hydroxy acids. The synergists are frequently metal deactivators. They have the ability to chelate metal ions which are prooxidants and which are responsible for the formation of free radicals in the initiation step of lipid autoxidation. There are two groups of synergists. One involves the action as a free radical acceptor and the other involves the action both as a free radical acceptor and a metal chelating agent (metal deactivator).

a) The first group of synergists, which acts only through accepting free radicals, regenerates the antioxidant from its free radical. Here, in the presence of the synergist (BH), the primary antioxidant (AH) is regenerated as follows.

ROO* (Lipid peroxy free radical)	+	AH (Antioxidant)	⟶	ROOH (Stable lipid)	+	A* (Antioxidant free radical)
A* (Antioxidant free radical)	+	BH (Synergist)	⟶	B* (Synergist free radical)	+	AH (Regenerated primary antioxidant)

b) The second group of synergists shows the above reaction and at the same time chelates the metal ions which otherwise would act as prooxidants through inducing the formation of free radicals in the initiation step.

IV. Effects of Lipid Autoxidation

Autoxidation of lipids has the following adverse effects on the quality of fish and fishery products.

1 Development of Rancid Flavour

The most important effect of lipid autoxidation is the development of a typical undesirable flavour called 'rancid flavour' or 'rancidity'. It is the development of this flavour for which autoxidation has been of great concern in fish processing. Basically, the development of rancidity can be attributed to two types of lipid deterioration such as 1) oxidative deterioration resulting in oxidative rancidity and 2) hydrolytic deterioration resulting in hydrolytic rancidity. Of the two types, oxidative deterioration is chiefly responsible for the development of rancidity in fish and fishery products.

Autoxidation of tissue lipids results in the development of oxidative rancidity whereas hydrolysis of lipids (into free fatty acids and glycerol) results in hydrolytic rancidity. Autoxidation of fish lipids produces primary, secondary and tertiary oxidation products. The primary products are the hydroperoxides (a type of peroxide) of lipids, which are mostly unstable, and do not have any remarkable effect on the flavour of fish. It is the secondary oxidation products, mostly carbonyl compounds, which cause rancidity. Therefore, organoleptic scores are more closely related to the secondary oxidation products. The most important secondary oxidation products are the carbonyls, particularly aldehydes, which are powerful flavour compounds and have very low flavour threshold. For example, the aldehyde, 2,4-decadienal has a flavour threshold of less than one parts per billion (ppb). That means, when present even at an extremely low concentration of one ppb, it can impart rancid flavour. It is the presence of double bond in the aldehydes, which generally is responsible for such low threshold. The rancid flavour becomes progressively more intense and unpleasant as lipid oxidation progresses.

2. Flavour Reversion

When lipids containing linolenic acid undergo oxidation, a typical unpleasant flavour, different from the rancid flavour, develops. This type of lipid deterioration is called flavour reversion. It requires considerably less oxygen than that required for autoxidation of lipids.

3. Discolouration

Besides the development of unpleasant flavour, autoxidation also results in discolouration of the lipids and imparts a yellow-brown or deep red colour, which is quite different from the original colour of the lipid or lipid containing tissue. Such yellow discolouration or browning is caused by ketones produced by autoxidation. However, the extent of discolouration depends upon the characteristics of the lipids and the degree of autoxidation. In case of extracted fish oils, sometimes pigments[8] are bleached by autoxidation with the result that the oils become light coloured. But, when this light-coloured oil is exposed to air for prolonged time, brown or red colour of a different origin develops. Besides, in fish oils, red colour also develops due to chemical combination of the aldehydes produced by autoxidation of highly unsaturated fatty acids and the TMA. Browning in frozen fish and darkening in canned fish may be attributed to the oxidation of the lipid components of the fish tissue.

4. Loss of Vitamin A

Liver of some fishes is a rich source of vitamin A. Autoxidation of lipids of the liver tissue destroys the vitamin A present in it to a large extent. Such destruction takes place due to the oxidation of vitamin A by the hydroperoxides produced by the autoxidation of tissue lipids.

5. Loss of Essential Fatty Acids

Due to the production of hydroperoxides, essential fatty acids are lost.

---

8 Pigments are the coloured lipid substances. Therefore, when lipid is extracted from any tissue, the pigments also get extracted into the extracted lipid dissolved in it. They impart a colour to the extracted lipid.

6. Loss of Nutritional Value of Proteins

The presence of some of the lipid oxidation products (keto hydroxy and epoxy compounds) result in loss of nutritive value of fish proteins. These oxidation products act on some amino acids such as cysteine, lysine, histidine and methionine resulting in loss of nutritional value of fish proteins. Such loss of nutritional value is evaluated through estimation of 'available lysine'.

7. Protein Denaturation

The oxidised tissue lipids react with tissue proteins. This reaction, called 'protein-lipid interaction' results in the formation of protein-lipid aggregates and consequently in the denaturation of proteins. The denaturation and aggregation of proteins results in the loss of their functional properties (solubility, water holding capacity, emulsifying capacity etc.), which is prominent particularly during cold storage of frozen fish. Due to loss of functional properties, the texture of fish meat deteriorates and it becomes tough, fibrous and chewy. Consequently the eating quality of fish gets deteriorated.

8. Development of Toxicity

Some of the lipid oxidation products (polymerised lipid peroxides) are toxic.

9. Non-enzymatic Browning (Maillard Reaction)

The oxidised lipids also react non-enzymatically with cytochrome c and haemoglobin to produce brown discolouration of fish tissues through a type of reaction called non-enzymatic browning or Maillard reaction.

10. Other Effects

Due to the autoxidation of lipids, the specific gravity, refractive index, viscosity, acid value and saponification value of fish oils increase whereas the iodine value and the content of ether insoluble bromide decrease.

V. Control of Autoxidation

Autoxidation of lipids has adverse effects on the quality of fish and fishery products. Therefore, it is always required to control or prevent autoxidation of lipids. It can be achieved by manipulating some of the factors, which influence the rate of autoxidation. The different methods by which autoxidation can be controlled or prevented are described below.

1. Lowering of Storage Temperature

Autoxidation of lipids is slowed down by storage of fish and fishery products at lower temperature. For every 10°C reduction in temperature, the rate of autoxidation falls by a factor of 2 to 3. Fatty fishes like herring, sardine and mackerel can be stored for more than six months without any remarkable rancid flavour of flesh by freezing them and storing in cold store at less than –18°C. Lean fishes like croaker and pink perch can be stored even longer. However, autoxidation is not completely inhibited even at the very low temperature of freezing and cold storage. It continues steadily throughout, though at a slower rate. That is why, development of rancidity by autoxidation of lipids is the major cause of quality deterioration of frozen fish.

2. Preventing the Exposure of Flesh

Autoxidation can also be controlled by preventing the exposure of flesh to atmospheric oxygen, thereby reducing access of oxygen to the lipids present in the fish tissue. It is achieved in different ways in different fishery products as given below.

A. Frozen Fish

In frozen fish, exposure of flesh to atmospheric oxygen is prevented in three different ways as given below.

(a) Pre-freezing Glazing

If fish is frozen as such and stored in cold store, it gets exposed to direct atmospheric air and to the air occluded between fishes. This leads to autoxidation and development of rancidity. To overcome this, fish or prawn are arranged in a tray and water is poured into it before freezing. This water fills the space between fishes and covers the fish completely. It has four effects. (1) It helps in getting a uniform block of fish. (2) It removes the air occluded among the fishes. (3) It acts as a barrier against direct contact with atmospheric air. (4) It prevents freeze dehydration. Among the four, the last three help in preventing autoxidation.

(b) Post-freezing Glazing

In case of individual frozen products such as individually quick frozen (IQF) shrimps or larger fishes frozen by air blast freezing, after freezing is over, the frozen product is given a 'post-freezing glazing treatment' by spraying chilled water to its surface or by giving it a quick dip treatment in chilled water. Due to the very low surface temperature of the product after freezing, a thin layer of water quickly solidifies on its surface and forms a coating on it. This prevents contact with atmospheric oxygen and thus prevents autoxidation. It can also be adopted in case of block frozen products.

(c) Proper Packaging

Frozen fish is packaged properly using packaging materials having good gas barrier properties, which prevent contact of the product with atmospheric oxygen. This prevents autoxidation of lipids.

B. Salted Fish

In salted fish, autoxidation continues at a faster rate due to its storage at normal room temperature as well as due to the prooxidant action of salt and its impurities. It can be overcome to some extent by allowing the salted fish to remain completely submerged in the brine. This prevents contact of fish with atmospheric oxygen. Moreover, the brine excludes the air occluded among the fishes. Such practice is adopted in the salting of fatty fishes, which are salted and sold as such without drying (*e.g.* in wet curing of sardine).

C. Dry Fish

In case of dry fish and salted-dried fish, lipid autoxidation is retarded by packing in airtight oxygen-impermeable packages, which prevent their exposure to atmospheric oxygen.

3. Reduction of Oxygen Concentration

Autoxidation can be prevented by removal of oxygen from airtight containers containing fish. In canning of fish, the oxygen concentration inside the tin can is reduced by complete removal of air during exhausting or vacuum seaming.

Another method adopted for extending the shelf life of chilled fresh fish is modified atmosphere packaging (MAP). Here, fresh fish is packed in an airtight container within which the atmosphere is modified desirably by flushing-in of required gases in required proportion. The three major gases used commercially are CO_2, N_2 and O_2. In case of lean fishes all the three gases are used in the ratio of 40 per cent: 30 per cent: 30 per cent:: CO_2: N_2:O_2. But in case of fatty fishes, O_2 is not used as it oxidises the tissue fats and turn them rancid. Here, the gas mixture consists of CO_2 and N_2 in the ratio of 40-60 per cent and 60-40 per cent respectively. Oxygen concentration in the pack is reduced by flushing these gases into the pack, which prevents autoxidation of tissue lipids.

4. Inhibition of Prooxidants

Prooxidants accelerate autoxidation of lipids. They include sunlight, ionising radiation, moisture, certain salts and their impurities. In case of fish oil, it should be prevented from exposure to sunlight and ionising radiations during storage. The oil should be made moisture-free. In case of salted fish, the salt should be free from impurities like iron and copper.

5. Use of Antioxidants

Natural antioxidants present in the fish meat as well as synthetic antioxidants incorporated to fish during processing slow down the process of autoxidation of lipids present in the fish tissue (Text box 8).

6. Use of Synergists

Besides antioxidants, there are certain substances called 'synergists', which are not highly effective by themselves as antioxidants, but when used together with antioxidants, increase the effectiveness of the antioxidants (Text box 9).

The overall description made above makes it clear that autoxidation can be controlled basically in four ways as follows:

1. Storage at low temperature
2. Exclusion of oxygen
3. Prevention of initiation and
4. Removal of free radicals

With respect to processed fish and fishery products, the most important is the control by storage at low temperature and the most effective is a combination of this process coupled with the exclusion of oxygen. The other methods are not so universally applicable, because prevention of initiation or removal of free radicals requires the introduction of controlling agents (antioxidants and synergists) to the lipid sites within the tissues and this causes problems. However, removal of free radical is applicable for extracted fish oils where the controlling agents can be mixed uniformly with the oils.

VI. Summary

1. What is Lipid Oxidation (Autoxidation) ?

Fish lipids are rich in unsaturated fatty acids. The unsaturation of the fatty acids is due to the presence of double bonds between carbon atoms in their hydrocarbon chains (Figure 51). When exposed to atmosphere, the atmospheric oxygen attacks and reacts with the double bonds on the fatty acids. The reaction of the atmospheric oxygen with the unsaturated double bonds of the fatty acids of lipid-molecules is called 'lipid oxidation' or 'autoxidation'. Autoxidation is not just simply addition of oxygen to the lipid molecules at the points of unsaturation; rather it is the complete set of reactions involving addition of oxygen to the lipid molecules as well as subsequent reactions. Lipid oxidation results in the formation of different types of end products, the presence of which in fish and fishery products imparts a typical rancid flavour called 'oxidative rancidity'.

2. Mechanisms of Lipid Oxidation

Lipid oxidation may occur in two ways, (1) Enzymatically by the enzyme lipoxygenase (formerly called lipoxidase) or (2) Non-enzymatically by autoxidation. However, non-enzymatic oxidation through autoxidation is more common than enzymatic oxidation. Therefore, autoxidation is of much greater concern in fish processing than enzymatic lipid oxidation. That is why, the terms 'lipid oxidation' and 'autoxidation' are used synonymously.

Lipid oxidation, which takes place through a non-enzymatic self-oxidation process, is called 'autoxidation' (*auto*: self; *oxidation*: reaction with oxygen). Autoxidation of lipids takes place through a 'free radical mechanism' as it starts and continues with the formation of free radicals. 'A free radical' is the component of a molecule, which has a free hand available due to removal of hydrogen radical from it. This free hand is free to bond with any reactive substance like oxygen. Lipid autoxidation is also a 'free radical chain reaction', because the product of the reaction of one molecule initiates the reaction of a second molecule, thus continuing the reaction in a chain.

Autoxidation of lipids takes place in four steps such as (I) Initiation, (II) Propagation, (III) Decomposition and (IV) Termination as described below.

I. Initiation

In this step one hydrogen radical (H*) is removed from an unsaturated fatty acid of a lipid molecule (R_1H) at the carbon atom next to the double bond (*i.e.* the α-carbon to the double bond) resulting in a lipid free radical (R_1*) (Figure 56). This reaction is brought about by agents like light and metal ions. These agents are called 'initiators' or 'chain initiators'. The free radical formed in the initiation step is stabilised by 'resonance'. That means, the radical is stabilised by the movement of the free hand from one atom to another (Figures 57 and 58). For example, it moves from C_8 to C_{10} and from C_{10} to C_8 (8 ↔ 10) or from C_9 to C_{11} and C_{11} to C_9 (9 ↔ 11) of the fatty acid.

II. Propagation

In this step, the few free radicals formed in the initiation step propagate to produce large number of free radicals (Figures 59 and 60). At first, the free radicals (R_1*) formed in the initiation step from the first lipid molecule (R_1H) combines with atmospheric oxygen (O_2) to form the first peroxy free radical (R_1OO*). Then this peroxy radical (R_1OO*) abstracts hydrogen radical (H*) from the unsaturated fatty acid of a second lipid molecule (R_2H) to produce a hydroperoxide (R_1OOH) and a new free radical (R_2*). While the hydroperoxide (R_1OOH), which is not a free radical is stable; the new free radical (R_2*) is highly reactive and combines with oxygen (O_2) to form the second peroxy free radical (R_2OO*). Then this peroxy radical (R_2OO*) abstracts hydrogen radical (H*) from the unsaturated fatty acid of a third lipid molecule (R_3H) to produce a hydroperoxide (R_2OOH) and a new free radical (R_3*). This reaction is repeated thousands of times and has the nature of a chain reaction.

III. Decomposition

The hydroperoxides formed during propagation are the primary oxidation products. They are generally not very stable. In the decomposition step, these feebly stable hydroperoxides are decomposed into 'secondary oxidation products' (Figure 65). The secondary oxidation products such as aldehydes are sometimes decomposed further to produce 'tertiary oxidation products' such as short chain fatty acids. Both, the secondary and tertiary oxidation products are stable. The hydroperoxides formed in the propagation step are decomposed in two steps as follows.

A. Decomposition of Hydroperoxide to Alkoxy Free Radical

In this step, lipid hydroperoxide is decomposed to an alkoxy free radical and a hydroxy free radical (Figures 66 and 67).

B. Decomposition of Alkoxy Free Radical to Aldehyde, Alcohol or Ketone

In this step of decomposition, the alkoxy free radical produced in the first step is decomposed in three ways to produce three products such as (a) Aldehyde, (b) Alcohol and (c) Ketone as described below.

(*a*) *Aldehyde Formation*: The alkoxy free radical may decompose to produce an aldehyde and a lipid free radical. The formation of the aldehyde from the alkoxy free radical involves the break down of the chain of fatty acid on the fat molecule. It may occur either on the left side or right side of the free radical. In this way two types of aldehydes may be produced (Figures 68 and 69).

(*b*) *Alcohol Formation*: The alkoxy radical may also abstract a hydrogen from another lipid molecule (R_2H) to yield an alcohol and a new free radical (R_2*) (Figure 70).

(*c*) *Ketone Formation*: The alkoxy free radical may react with a new free radical (R_2*) formed in the above step to terminate the chain reaction with formation of ketone (Figure 71).

The decomposition products such as aldehydes, alcohols and ketones impart rancid flavour to the oxidised lipids. Among them, the aldehydes impart the highest rancid flavour. If the aldehyde has double bond, the rancid flavour is still more. The aldehydes can be further oxidised to 'tertiary oxidation products' such as carboxylic acids like free fatty acids or other tertiary oxidation products (Figure 65).

IV. Termination

The chain reaction of autoxidation does not continue indefinitely. It is stopped or terminated in the termination step, which is the last step of the autoxidation process. In this step, the free radicals react within themselves to yield non-reactive products as follows.

A. A pair of free radicals may react to form a stable compound.

$$R_1^* + R_2^* \longrightarrow R_1 - R_2$$

B. A free radical may react with a peroxy radical produced in the first phase of propagation step to form a stable compound.

$$R_1^* + R_2OO^* \longrightarrow R_1OOR_2$$

C. A pair of peroxy free radicals may react to form a stable compound.

$$R_1OO^* + R_2OO^* \longrightarrow R_1OOR_2OO$$

D. A large number (n) of peroxy free radicals may react to form a stable product.

$$n\,(ROO^*) \longrightarrow (ROO)_n$$

3. Factors Influencing Lipid Autoxidation

Autoxidation of lipids is influenced by the following factors.

(*a*) Composition of the Lipids

Composition of tissue lipids of fish with respect to degree of unsaturation (*i.e.* number of double bonds in the fatty acid residues), geometry of double bonds, position of the double bonds, disposition of the fatty acid residues within the triglycerides molecule and the freedom of fatty acid residues have great influence on rate of autoxidation.

(*b*) Temperature

Temperature has marked influence on the rate of lipid autoxidation. The rate of autoxidation increases with increasing temperature and decreases with decreasing temperature usually by a factor of 2 to 3 for every 10°C change in temperature. However, unlike microbial activity, which is completely inhibited below a temperature of –10°C, autoxidation is not completely inhibited even at very low temperatures of freezing and cold storage of fish. Therefore, oxidative deterioration is the most important agent of quality loss in frozen fish during cold storage.

(*c*) Degree of Exposure of Flesh

Autoxidation takes place at a faster rate if the tissue lipids are more exposed to atmosphere. As oxygen diffuses very slowly through the fish tissue, the lipids, which are embedded deep in the tissue unexposed to atmospheric oxygen get oxidised very slowly.

(*d*) Oxygen Concentration

If oxygen supply is unlimited, the rate of autoxidation does not depend upon oxygen concentration. But when oxygen concentration is very low, the rate of autoxidation is proportional to the concentration of oxygen.

(*e*) Fat Content of Fish, its Types and Disposition

Fatty fishes oxidise faster than lean fishes; triglycerides oxidise faster than phospholipids and depot fats oxidise faster than membrane bound lipids.

(*f*) Location of Lipids

Lipids of dark lateral muscle of fish tend to oxidise faster than those of the white muscle.

(g) Extraction

Lipids, when present in fish carcass, undergo rapid autoxidation than when extracted as fish oil.

(h) Presence of Prooxidants

Certain substances are known to accelerate the rate of autoxidation. They are called 'prooxidants'. The presence or addition of prooxidants to fish leads to rapid autoxidation. Some of the important prooxidants are light and ionising radiations, metals, metallic soaps, common salt, salt impurities, moisture, heme compounds, flesh components and lipoxygenase.

(i) Presence of Antioxidants

Many foods contain natural antioxidants (Text box 8). Among them tocopherols (vitamin E) are the most important. They are present in large quantity in vegetable oils than in animal fats for which vegetable oils are more stable (do not develop rancidity) than animal fats. When oil is extracted from the fatty tissues, say muscle or liver of fish, some amount of natural antioxidants present in these tissues gets extracted into the extracted crude oil. But when the crude oil is subjected to the refining process, the natural antioxidants are lost. Hence, such refined oils are more susceptible to autoxidation (more unstable) than the crude oils. Squalene present in fish lipids is also a natural antioxidant. To prevent autoxidation of tissue lipids during frozen storage of fish, synthetic commercial antioxidants are added to it. It is important to note that, at high concentration, antioxidants may have a prooxidants effect. Therefore, addition of antioxidants beyond optimum amount may result in enhancement of lipid autoxidation.

(j) Presence of Synergists

The presence of synergists in lipid tissues or extracted oils enhances the prevention of autoxidation by the added antioxidants (Text box 9).

(k) Cold Storage Dehydration

Dehydration of frozen fish during cold storage results in direct exposure of fish tissue to atmospheric oxygen, which accelerates autoxidation.

4. Effects of Lipid Autoxidation

Autoxidation of lipids has the following adverse effects on the quality of fish and fishery products.

(a) Development of Rancid Flavour

Basically, the development of rancidity can be attributed to two types of lipid deterioration such as 1) oxidative deterioration resulting in oxidative rancidity and 2) hydrolytic deterioration resulting in hydrolytic rancidity. Of the two types, oxidative deterioration is chiefly responsible for the development of rancidity in fish and fishery products.

(b) Flavour Reversion

When lipids containing linolenic acid undergo oxidation, a typical unpleasant flavour, different from the rancid flavour, develops. This type of lipid deterioration is called flavour reversion. It requires considerably less oxygen than that required for autoxidation of lipids.

(c) Discolouration

Autoxidation also results in discolouration of the lipids and imparts a yellow-brown or deep red colour, which is quite different from the original colour of the lipid or lipid containing tissue. Such yellow discolouration or browning is caused by ketones produced by autoxidation.

(d) Loss of Vitamin A

Autoxidation of lipids of the liver tissues destroys the vitamin A present in it to a considerable extent.

(*e*) Loss of Essential Fatty Acids

Due to the production of hydroperoxides, essential fatty acids are lost.

(*f*) Loss of Nutritional Value of Proteins

The presence of some of the lipid oxidation products (keto hydroxy and epoxy compounds) result in loss of nutritive value of fish proteins.

(*g*) Protein Denaturation

The oxidised tissue lipids react with tissue proteins. This reaction, called 'protein-lipid interaction' results in the formation of protein-lipid aggregates and consequently in the denaturation of proteins. The denaturation and aggregation of proteins results in the loss of their functional properties (solubility, water holding capacity, emulsifying capacity etc.), which is prominent particularly during cold storage of frozen fish. Due to loss of functional properties, the texture of fish meat deteriorates and it becomes tough, fibrous and chewy. Consequently the eating quality of fish gets deteriorated.

(*h*) Development of Toxicity

Some of the lipid oxidation products (polymerised lipid peroxides) are toxic.

(*i*) Non-enzymatic Browning (Maillard Reaction)

The oxidised lipids also react non-enzymatically with cytochrome c and haemoglobin to produce brown discolouration of fish tissues through a type of reaction called non-enzymatic browning or Maillard reaction.

(*j*) Other Effects

Due to the autoxidation of lipids, the specific gravity, refractive index, viscosity, acid value and saponification value of fish oils increase whereas the iodine value and the content of ether insoluble bromide decrease.

5. Control of Autoxidation

Autoxidation of lipids has adverse effects on the quality of fish and fishery products. Therefore, it is always required to control or prevent autoxidation of lipids. The different methods by which autoxidation can be controlled or prevented are described below.

(*a*) Lowering of Storage Temperature

Autoxidation of lipids is slowed down by storage of fish and fishery products at lower temperature. For every 10°C fall in temperature, the rate of autoxidation falls by a factor of 2 to 3.

(*b*) Preventing the Exposure of Flesh

Autoxidation can also be controlled by preventing the exposure of flesh to atmospheric oxygen, thereby reducing access of oxygen to the lipids present in the fish tissue. It is achieved in different ways in different fishery products as given below.

(*i*) *Frozen Fish*

In frozen fish, exposure of flesh to atmospheric oxygen is prevented in three different ways as given below.

(*a*) Pre-freezing glazing

(*b*) Post-freezing glazing

(*c*) Proper packaging

(*ii*) *Salted Fish*

In salted fish, autoxidation continues at a faster rate due to its storage at normal room temperature as well as due to the prooxidant action of salt and its impurities. It can be overcome to some extent by allowing the salted fish to remain completely submerged in the brine.

(iii) Dry Fish

In case of dry fish and salted-dried fish, lipid autoxidation is retarded by packing in airtight oxygen-impermeable packages, which prevent their exposure to atmospheric oxygen.

(c) Reduction of Oxygen Concentration

Autoxidation can be prevented by removal of oxygen from airtight containers containing fish as in case of canned fish and to some extent in fresh chilled fish under modified atmosphere packaging (MAP).

(d) Inhibition of Prooxidants

Prooxidants such as sunlight, ionising radiation, moisture, certain salts and their impurities accelerate autoxidation of lipids. Therefore, extracted fish oil should be made moisture-free. In case of salted fish, the salt should be free from impurities like iron and copper.

(e) Use of Antioxidants

Natural antioxidants present in the fish meat as well as synthetic antioxidants incorporated to fish during processing slow down the process of autoxidation of lipids present in the fish tissue (Text box 8).

(f) Use of Synergists

Besides antioxidants, there are certain substances called 'synergists', which are not highly effective by themselves as antioxidants, but when used together with antioxidants, increase the effectiveness of the antioxidants (Text box 9).

The overall description made above makes it clear that autoxidation can be controlled basically in four ways as follows.

1. Storage at low temperature
2. Exclusion of oxygen
3. Prevention of initiation and
4. Removal of free radicals

With respect to processed fish and fishery products, the most important is the control by storage at low temperature and the most effective is a combination of this process coupled with the exclusion of oxygen. The other methods are not so universally applicable, because prevention of initiation or removal of free radicals requires the introduction of controlling agents (antioxidants and synergists) to the lipid sites within the tissues and this causes problems. However, removal of free radical is applicable for extracted fish oils where the controlling agents can be mixed uniformly with the oils.

7

Discolouration

After death of fish several changes occur in the carcass. Some of them result in the discolouration of the meat. Discolouration is any change in the original colour of fish after death. Post mortem discolouration can be grouped into the following categories.

I. Black spot formation in shrimp (Melanosis)

II. Brown discolouration of red meat of fish

III. Red/brown discolouration of white fish fillets

I. Black Spot Formation in Shrimp (Melanosis)

The quality problem, very often encountered in the post harvest storage of shrimps and other shellfishes by icing, is the formation of black spots on their shells followed by a general blackening of the shell and meat. Such discolouration is called 'black spot formation' or 'melanosis'.

A. Chemistry of Melanosis

The free amino acid, tyrosine, present in the muscle of shrimps and other shellfishes, in presence of the oxygen in the air, is oxidised by the endogenous tissue enzyme tyrosinase to insoluble black or black brown pigments called 'melanin'[1]. (Figure 73). Accumulation of the melanin pigments in the shell and meat results in the formation of black spots.

$$\text{Free tyrosine in muscle} \xrightarrow[\text{Tyrosinase of tissue}]{[O_2]} \text{Melanin} \longrightarrow \text{Black spot}$$

Figure 73: Mechanism of black spot formation

Melanosis is caused by the enzymatic formation of precursor compounds, which further undergo condensation spontaneously and/or react with cellular constituents to form melanin. Melanin is a polymer of tyrosine catabolites. It is formed from tyrosine in the following steps (Figure 74).

1. Small quantity of tyrosine is present as free amino acid in the muscle of shrimps and other shellfishes
2. It is also formed in the muscle from free phenylalanine.

---

1 Melanin is also the pigment, which imparts black colour to hair and skin of human beings.

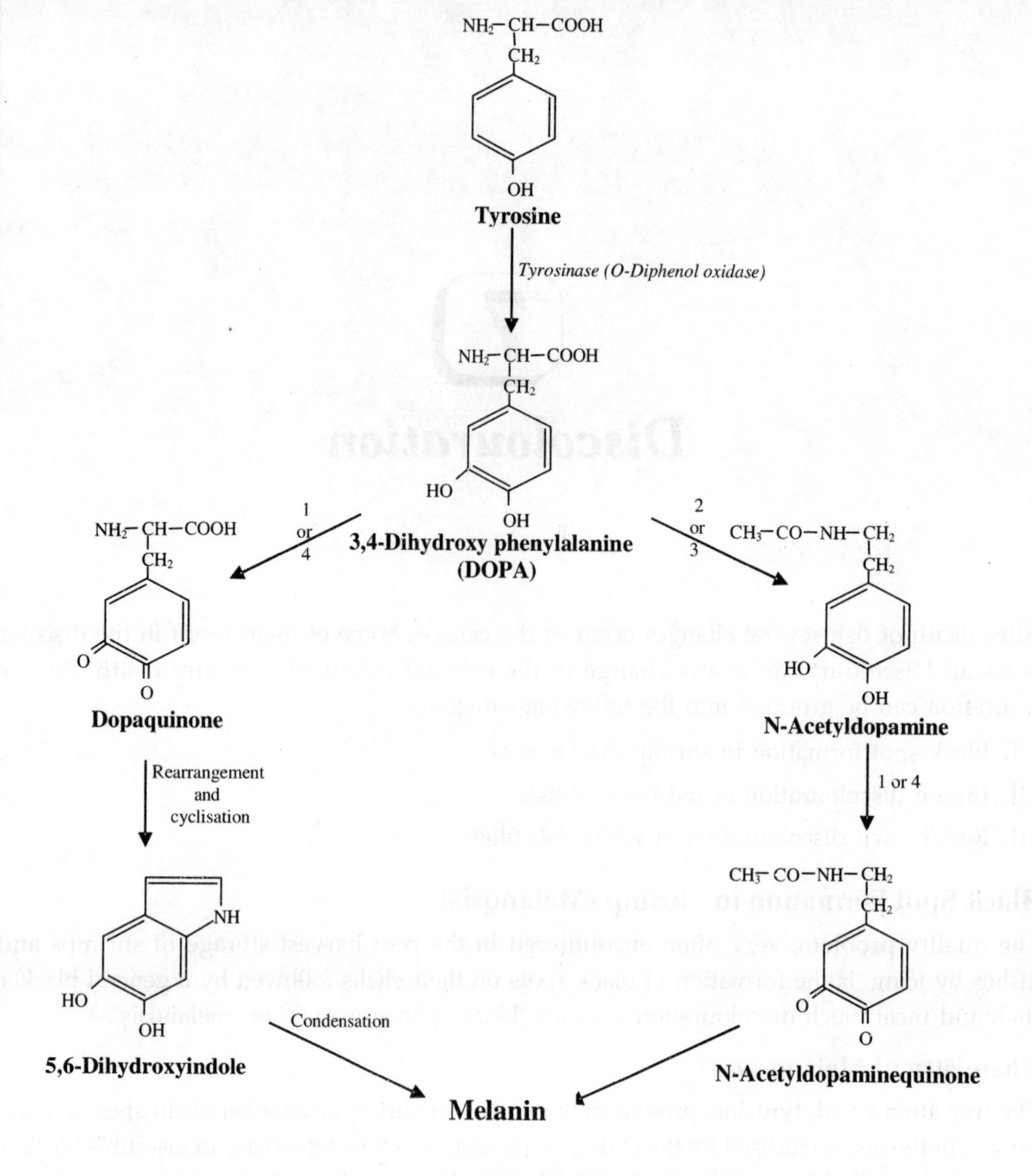

1: DOPA oxidase *2: DOPA dicarboxylase* *3: Transacetylase* *4: Nonenzymatic autoxidation catalysed by light*

Figure 74: Post mortem formation of melanin in shrimp

3. The endogenous muscle enzyme, tyrosinase, (o-diphenol oxidase) catalyses the oxidation of tyrosine to 3,4-dihydroxy phenylalanine (DOPA).
4. While the monohydric phenol tyrosine is fairly stable to chemical oxidation, the dihydric phenol such as DOPA is very unstable and readily undergoes oxidation.
5. DOPA is oxidised to dopaquinone either enzymatically by DOPA oxidase or non-enzymatically by autoxidation catalysed by light.
6. Dopaquinone undergoes rearrangement and cyclises to form 5,6-dihydroxyindole.
7. 5,6-Dihydroxyindole condenses to produce the pigment melanin.

5a. *In another path,* DOPA is enzymatically converted to N-acetyldopamine by dopadicarboxylase or transacetylase.

6a. N-acetyldopamine is oxidised to N-acetyldopaminequinone either enzymatically by DOPA oxidase or non-enzymatically by autoxidation catalysed by light.

7a. N-acetyldopamine quinone changes to melanin.

All these reactions are mediated by a number of enzymes collectively known as 'phenolases'. The enzymes remain active throughout post harvest processing unless shrimp is frozen or cooked.

Both enzymatic and non-enzymatic oxidation is responsible in post harvest melanin formation. Melanin precursors and phenolases are found in high concentration in cephalothorax as well as in the tail of shrimps. The melanin formed from the precursors readily condenses with amino groups on proteins in an irreversible tanning process. Once the pigment is formed, it cannot be removed without the destruction of proteins. If shrimp is iced after harvesting, the washing action of melting ice removes many water-soluble compounds including free tyrosine and phenylalanine, which are the mother substances of melanin.

B. Factors Affecting Rate of Melanosis

Several factors affect the rate of melanosis in shrimps. They are as follows.

1. Species

Different species of shrimps show different rates of melanosis due to difference in the activity of their phenolases. *Penaeus indicus, Metapenaeus affinis* and *M. monoceros* show higher phenolase activity than *Parapenaeopsis stylifera* and *M. dobsoni. P. indicus* is more prone to melanosis than *Metapenaeus affinis, M. dobsoni* and *M. monoceros.*

2. Habitat

Shrimps harvested from habitats rich in organic matter show melanosis more readily. Shrimps harvested from Indian Ocean blacken at a rate different from those harvested from Arabian Sea. This may be due to the difference in the nutrient content of the two habitats.

3. Size of Shrimps

Larger shrimps are more susceptible to melanosis than smaller ones.

4. Phenolase Activity

Different species of shrimp show different phenolase activity. Species showing higher phenolase activity undergo melanosis at faster rates.

5. Temperature

Rate of melanosis increases with increase in temperature (upto 55°C) as the phenolase activity increases with increase in temperature. As the temperature of storage is lowered, the rate of melanosis decreases. This is the reason prawn should be iced immediately after harvesting to prevent melanosis.

6. Exposure to Air

Exposure to air (oxygen) enhances the rate of melanosis. This is because, both tyrosine and other melanin precursors as well as the enzyme phenolase are present in the body of shrimps. It is only the atmospheric oxygen, which is required to complete the process of melanosis. Shrimps held at 0°C without ice come in direct contact with air due to which they show melanosis more quickly than those held at 0°C in ice. The melt water also washes down many water-soluble compounds including tyrosine, phenylalanine and other precursors thereby slowing down melanosis.

C. Control of Melanosis

Melanosis can be controlled in the following ways.

1. Beheading the Shrimp

Melanosis can be minimised by beheading the shrimps as soon as possible after harvesting. This is because, the phenolase enzymes are found in higher concentration in the head (cephalothorax) than in the tail (abdomen). However, melanosis cannot be completely avoided by this method.

2. Icing

A simple and effective way of preventing melanosis is to pack the shrimps with ice immediately after harvesting and completely cover them with water in the containers so as to prevent oxygen in the air from coming in contact with the shrimps. Icing not only retards the general spoilage of shrimps, it also prevents melanosis to a greater extent by slowing down the chemical reactions involved in melanosis. Therefore, the practice of keeping freshly caught shrimp in buckets for some hours prior to transferring to icing facilities not only accelerates general spoilage, but also is conducive to the development of black spots.

3. Chemical Treatment

To inhibit melanosis effectively, shrimps are subjected to chemical treatment prior to icing. The following chemicals are used commercially to prevent melanosis.

(a) Sodium Bisulphite

Sulphites have been used to control melanosis since 1950s.The most commonly used sulphites are sodium bisulphite ($NaHSO_3$), sodium metabisulphite ($Na_2S_2O_3$) and sodium sulphite (Na_2SO_3). Icing after a dip treatment of shrimps in dilute solution of sodium bisulphite or icing using ice incorporated with sodium bisulphite is generally practised in commerce to control melanosis. In many countries, the dip treatment is employed on the fishing vessel itself followed by icing or brine freezing of the sulphite-treated shrimp. The USA regulations allow a 1-minute dip treatment in a 1.25 per cent solution of sodium metabisulphite so that the treated shrimp has a maximum of 100 ppm of metabisulphite in the meat. Use of sulphites to control melanosis in shrimps has created several problems. (1) Sulphites cannot prevent melanosis very effectively. (2) Sulphites have life-threatening hazards on persons suffering from asthma. (3) The sulphur dioxide gas produced from the sulphites accumulates in the fish hold of the fishing vessels and is extremely lethal. Several deaths of fishermen who were exposed to the sulphur dioxide in the fish hold have been recorded.

(b) 4-Hexylresorcinol

With increased regulations on food additives and consumers' awareness about the risks associated with sulphite-treated foods, it became imperative to search for safe and effective alternatives to sulphites for control of melanosis in shrimps. One group of chemicals namely the '4-substituted resorcinols' was found to inhibit melanosis effectively. Among them, 4-hexylresorcinol is the most effective in inhibiting melanosis. Due to its use in human consumables since 1920s, extensive toxicological studies have been undertaken which have recommended it as Generally Recognised As Safe (GRAS[2]). To control melanosis effectively and safely, immediately after harvesting, shrimp is given a 1-minute dip treatment in 50 ppm (50 mg/l) solution of 4-hexylresorcinol which gives a residual level of 0.7-1.0 ppm in the meat. After dip treatment, the shrimp is iced immediately.

(c) Other Chemicals

Few other chemicals and chemical mixtures, which have been reported to inhibit melanosis, but have not found commercial application till now include (1) sulphur containing amino-carboxylic acids, (2) tyrosine, (3) diethyldithiocarbamate, (4) a mixture of sodium bisulphite, sodium chloride and dextrose, (5) a mixture of sodium metabisulphite, citric acid, sodium bicarbonate and permitted antioxidants.

---

2 GRAS (Generally Recognised As Safe) is a list of natural and synthetic substances, which after thorough investigations have been regarded by the FDA (Food and Drug Administration) of the USA, as safe for use in food and drug within certain prescribed limits.

II. Brown Discolouration of Red Meat of Fish

• In terrestrial animals, the red muscle fibers and white muscle fibers are mixed within each muscle thereby constituting the homogenous red meat. On the contrary, the bulk of the muscle in most fishes consists of white muscle fibers alone, which constitutes their white meat. However, the muscle of most of the pelagic fishes like mackerel, sardine, herring and tuna contain considerable amount of red fibers besides the white fibers. Unlike in terrestrial animals, in these fishes, the red and white fibers are arranged in discrete zones forming the white muscle (containing white fibers) and dark muscle (containing red fibers). While the white muscle constitutes the bulk of the musculature, the dark muscle is found beneath the lateral line region superficially running roughly parallel to the long axis of the body. The dark colour of dark muscle is mainly due to its high content of the red muscle pigment 'myoglobin'.

Myoglobin is the red coloured muscle pigment (Figure 75). It is a globular protein present in the muscle, which imparts red colour to meat. It is a complex protein as it has two parts (moiety) as follows.

1. Globin (Protein moiety)
2. Heme (Non-protein moiety)

1. Globin

It is the protein moiety of myoglobin. It is a globular protein with molecular weight of 16,000-17,000. Approximately 75 per cent of the amino acid residues of the globin molecule are present in 8 right-handed alpha-helices. Each helix has 7-20 amino acid residues. Starting from the N-terminal, these 8 helices are named helices A through H. Individual amino acid residues are designated by a letter for the helix in which they remain and a number that indicates their distance from the N-terminal of that helix. For example, "His F8" refers to the eighth residue from the N terminal in the F helix and identifies it as histidine.

2. Heme

It is the non-protein moiety of myoglobin. It is composed of two parts as follows.

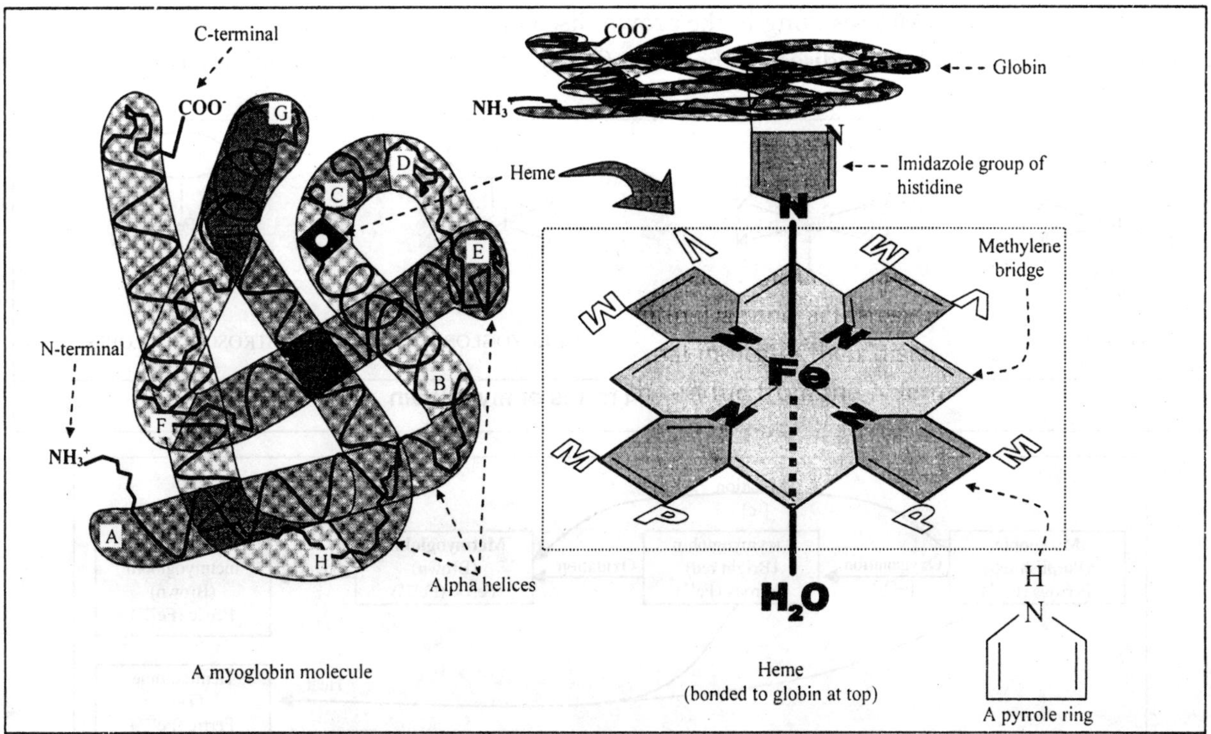

Figure 75: Structure of myoglobin (globin + heme)

(*a*) Porphyrin Ring

It is a large planar ring, which is made up of four sub-units called pyrrole (deep gray in figure). Pyrrole is a heterocyclic compound. The four pyrrole rings are linked together by methylene bridges. An iron atom remains in the center of the porphyrin ring. The four pyrrole rings, the methylene bridges and the iron atom are coplanar (remain in one plain, shown as light gray in the figure). There are three different kinds of side chains-methyl (M), vinyl (V), and propyl (P). This ring structure can form stable complexes with a large number of metal ions. The complex with iron is known as heme. When heme is bound to globins through the side chains, the resulting compounds are myoglobin and hemoglobin. The main difference between hemoglobin and myoglobin is that hemoglobin has four polypeptide chains (tetrameric) while myoglobin has only one (monomeric).

(*b*) An Iron Atom

It remains at the center of the heme. It has six bonds, which represent six bond orbitals in which the iron is sharing two electrons with another atom. In this case, the iron has not contributed any electron itself but has accepted six pairs from some other atoms-five pairs from nitrogen atoms, and one pair from oxygen. Four of the nitrogen atoms are in the porphyrin ring, and one is in the imidazole group of a histidine molecule in the amino acid chain of the globin. It is the sixth bond orbital, which provides for the function of the molecule in that this orbital is available for the complexing of any atom having an electron pair to donate. The fifth and sixth coordination positions of Fe^{2+} are directed perpendicular to and directly above or below the plane of the heme ring.

A. Chemistry of Brown Discolouration

The white muscle of few fishes like tuna is red similar to the red meat of terrestrial animals. This is due to the high content of myoglobin in the white muscle. Such red meat undergoes discolouration to the undesirable brown colour similar to the discolouration of red meat of terrestrial animals. The brown discolouration is due to the change in the myoglobin. When tuna is deskinned or cut into pieces, the meat comes in contact with atmospheric oxygen. In the presence of oxygen, the purplish red coloured myoglobin (Mb) of meat is 'oxygenated' to produce the bright red coloured oxymyoglobin (MbO_2), characteristic of fresh tuna loins (Figures 76 and 77). Subsequently, both Mb and MbO_2 are 'oxidised' to brown metmyoglobin (MMb) resulting in the brown discolouration of meat. In Mb and MbO_2, iron is in ferrous (Fe^{++}) state, which is oxidised to the ferric (Fe^{+++}) state in MMb. If metmyoglobin is heated, its

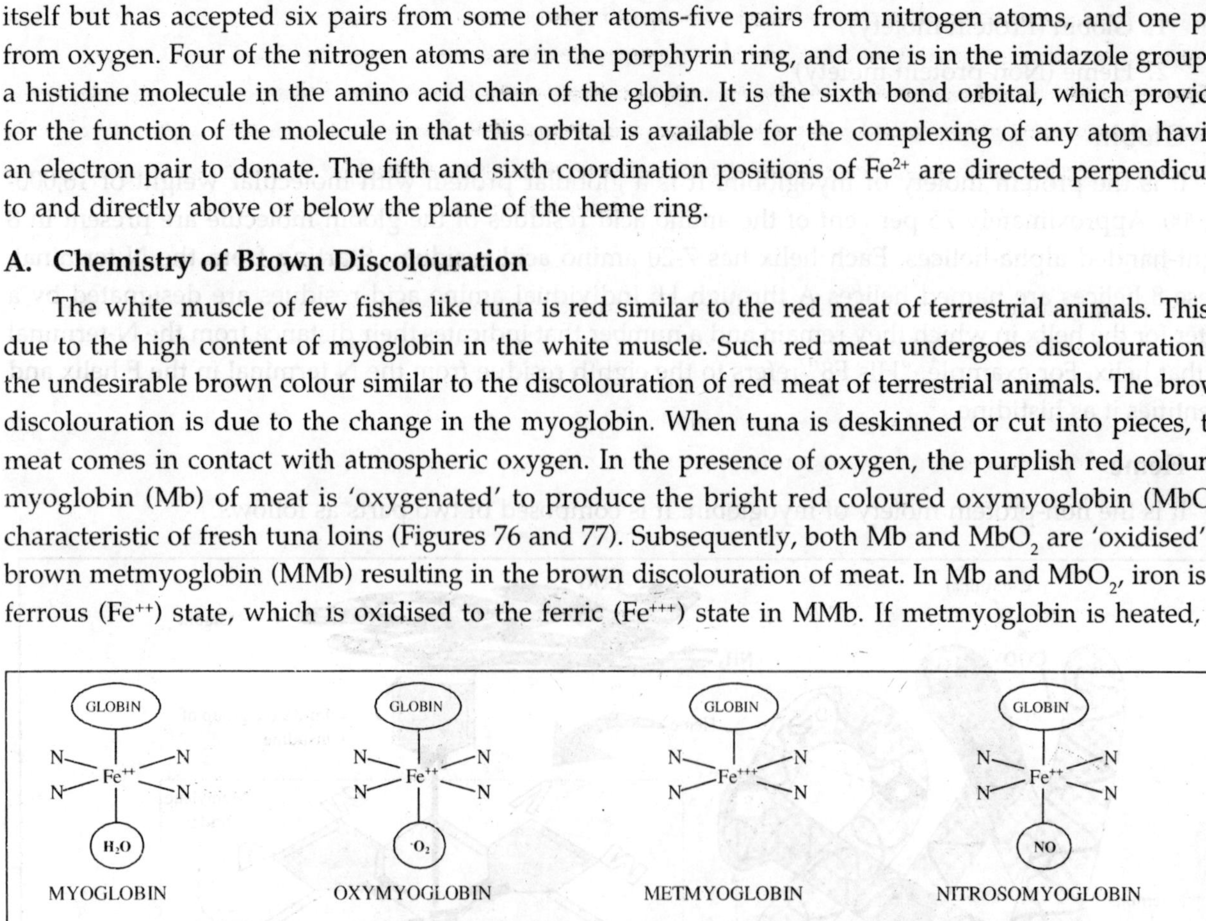

Figure 76: Different forms of myoglobin

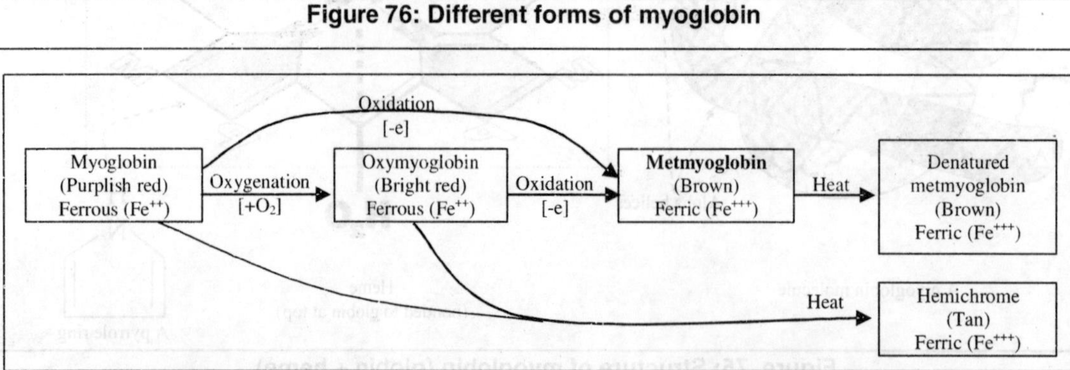

Figure 77: Chemistry of brown discolouration of red meat of tuna

globin moiety gets denatured thereby producing denatured metmyoglobin, which is also brown in colour.

If fresh meat is heated, the protein portion of MbO_2 (*i.e.* globin) gets denatured and the heme with iron in ferrous state (Fe^{++}) changes to hemichrome with iron in ferric state (Fe^{+++}). The hemichrome is responsible for the tan colour of fresh cooked tuna. In presence of reducing agents, hemichrome with iron in ferric state (Fe^{+++}) changes to hemochrome with iron in ferrous state (Fe^{++}). The hemochrome has pink colour and is responsible for the pink colour of fresh tuna cooked in reducing condition.

B. Control of Brown Discolouration

The brown discolouration of red meat of tuna is controlled by 'colour fixation' through curing with sodium nitrite ($NaNO_2$). This takes place in the following steps.

1. Sodium nitrite dissolves in water and dissociates to sodium cation (Na^+) and nitrite anion (NO_2^-) (Figures 78 and 79).
2. The nitrite anion combines with water to produce nitrous acid (HNO_2) and nitric oxide (NO).
3. Nitric oxide (NO) penetrates into the meat and binds with myoglobin to produce metastable nitric oxide-myoglobin (NOMb). It is also called nitrosomyoglobin. Actually NO binds to the Fe atom in the center of the heme of myoglobin thereby preventing its oxidation to brown metmyoglobin (MMb).
4. Cooking changes metastable nitrosomyoglobin to stable nitrosohemochrome (or nitrosylhemochrome), which has the desirable stable pink colour. Due to cooking, the globin moiety of myoglobin gets denatured. Thus, nitrosohemochrome consists of denatured globin and heme pigment with NO attached to it.

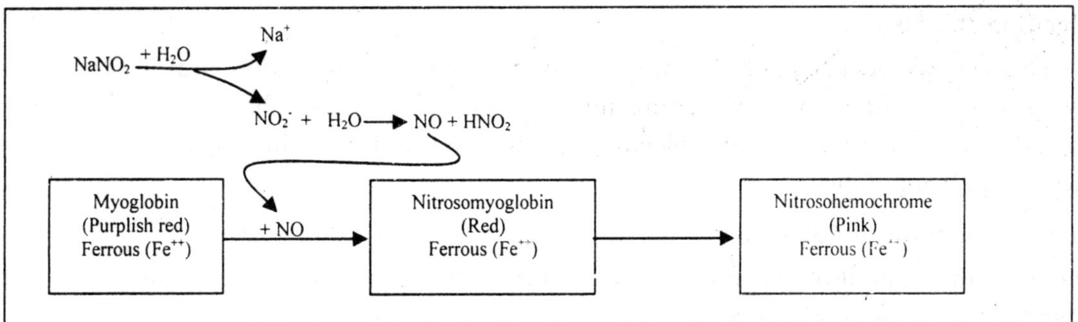

Figure 78: Colour fixation by curing

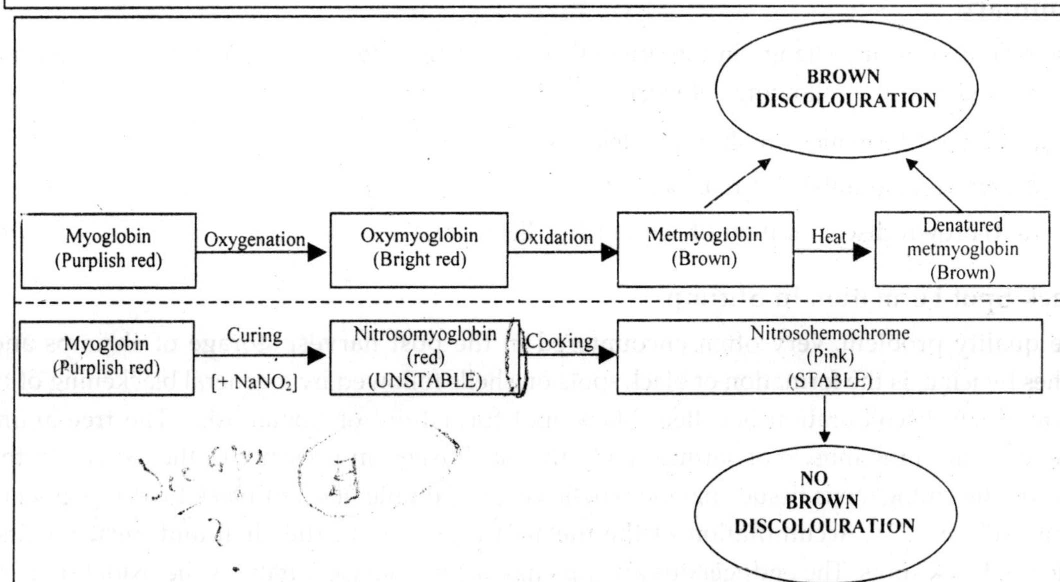

Figure 79: Brown discolouration of red meat and its control (summary)

III. Red/Brown Discolouration of White Fish Fillets

Fillets are the pieces of fish flesh obtained by cutting a fish longitudinally along both the sides of the backbone. Thus each fish gives two fillets. The desirable colour of fillets from white fishes (lean, demersal fishes) is white. Sometimes, the colour of fillets changes from original white to red or brown colour. This is called red or brown discolouration of white fish fillets.

A. Causes of Red or Brown Discolouration of White Fish Fillets

The reasons for the red or brown discolouration of white fish fillets are given below.

(*a*) No Bleeding

If fish is filleted immediately after death without bleeding it, fillets undergo undesirable red or brown discolouration. This occurs because blood remains in the flesh and clots there imparting brown colour to the originally white flesh.

(*b*) Rough Handling

Discolouration of fillets may also result from rough handling during harvesting and handling the catch while the fish is still alive. Physical mishandling in nets (long trawling time, very large catch) or on the deck (stepping on the fish or throwing boxes, containers and other items on top of fish) may cause bruises, rupture of blood vessels and blood may ooze into the flesh (haematoma). However, heavy pressure on dead fish, when blood has already clotted (overloading of fish boxes) does not enhance discolouration.

B. Control of Red or Brown Discolouration of White Fish Fillets

The red or brown discolouration of fillets can be controlled by the following ways.

(*a*) Bleeding the Fish

If fish is subjected to bleeding immediately after harvesting, blood drains off the flesh and discolouration does not occur. For bleeding, immediately after the fish is harvested, the throat is cut on a grate from which chutes lead to a bleeding bin where the fish is allowed to bleed for at least 15 minutes before being gutted.

(*b*) Proper Handling of Catch

Discolouration can also be controlled by careful handling of fish during harvesting and handling the catch while the fish is still alive.

IV. Summary

Discolouration is any change in the original colour of fish after death. Post mortem discolouration can be grouped into the following categories.

1. Black spot formation in shrimp (Melanosis)
2. Brown discolouration of red meat of fish
3. Red/brown discolouration of white fish fillets

1. Black Spot Formation in Shrimp

The quality problem, very often encountered in the post harvest storage of shrimps and other shellfishes by icing, is the formation of black spots on shells followed by a general blackening of the shell and meat. Such discolouration is called 'black spot formation' or 'melanosis'. The free amino acid, tyrosine, present in the muscle of shrimps and other shellfishes, in presence of the oxygen in the air, is oxidised by the endogenous tissue enzyme tyrosinase to insoluble black or black brown pigments called 'melanin'. (Figure 73). Accumulation of the melanin pigments in the shell and meat results in the formation of black spots. The endogenous shrimp enzyme, tyrosinase, catalyses the oxidation of tyrosine to 3,4-dihydroxy phenylalanine (DOPA) which is very unstable and is readily oxidised to dopaquinone

(Figure 74). Dopaquinone undergoes rearrangement and cyclises to form 5,6-dihydroxyindole, which condenses to produce the pigment melanin. In another path, DOPA is enzymatically converted to N-acetyldopamine, which is further oxidised to N-acetyldopaminequinone. Finally, N-acetyldopamine quinone changes to melanin. All these reactions are mediated by a number of enzymes collectively known as 'phenolases'. The enzymes remain active throughout post harvest processing unless shrimp is frozen or cooked. Melanin precursors and phenolases are found in high concentration in cephalothorax as well as in the tail of shrimps. The melanin formed from the precursors readily condenses with amino groups on proteins in an irreversible tanning process. Once the pigment is formed, it cannot be removed without the destruction of proteins. If shrimp is iced after harvesting, the washing action of melting ice removes many water-soluble compounds including free tyrosine and phenylalanine, which are the mother substances of melanin.

Several factors affect the rate of melanosis in shrimps. They include species of shrimp, its habitat, size, phenolase activity, storage temperature and exposure of the shrimp to air.

Melanosis can be controlled by beheading the shrimps and icing it immediately after harvesting. Chemical treatments before icing to control melanosis include a 1-minute dip treatment in a 1.25 per cent solution of sodium metabisulphite or a 1-minute dip treatment in a 50 ppm (50 mg/l) solution of 4-hexylresorcinol.

2. Brown Discolouration of Red Meat of Fish

The white muscle of few fishes like tuna is red similar to the red meat of terrestrial animals. This is due to the high content of myoglobin in the white muscle. Such red meat undergoes discolouration to the undesirable brown colour similar to the discolouration of red meat of terrestrial animals. The brown discolouration is due to the change in the myoglobin. When tuna is deskinned or cut into pieces, the meat comes in contact with atmospheric oxygen. In the presence of oxygen, the purplish red coloured myoglobin (Mb) of meat is 'oxygenated' to produce the bright red coloured oxymyoglobin (MbO_2), characteristic of fresh tuna loins (Figures 76 & 77). Subsequently, both Mb and MbO_2 are 'oxidised' to brown metmyoglobin (MMb) resulting in the brown discolouration of meat. If metmyoglobin is heated, its globin moiety gets denatured thereby producing denatured metmyoglobin, which is also brown in colour.

The brown discolouration of red meat of tuna is controlled by 'colour fixation' through curing with sodium nitrite ($NaNO_2$). Sodium nitrite dissolves in water and dissociates to sodium cation (Na^+) and nitrite anion (NO_2^-) (Figures 78 & 79). The nitrite anion combines with water to produce nitrous acid (HNO_2) and nitric oxide (NO). Nitric oxide (NO) penetrates into the meat and binds with myoglobin to produce metastable nitric oxide-myoglobin (NOMb). It is also called nitrosomyoglobin. Actually NO binds to the Fe atom in the center of the heme of myoglobin thereby preventing its oxidation to brown metmyoglobin (MMb). Cooking changes metastable nitrosomyoglobin to stable nitrosohemochrome (or nitrosylhemochrome), which has the desirable stable pink colour. Due to cooking, the globin moiety of myoglobin gets denatured. Thus, nitrosohemochrome consists of denatured globin and heme pigment with NO attached to it.

3. Red/Brown Discolouration of White Fish Fillets

Fillets are the pieces of fish flesh obtained by cutting a fish longitudinally along both the sides of the backbone. Thus each fish gives two fillets. The desirable colour of fillets from white fishes (lean, demersal fishes) is white. Sometimes, the colour of fillets changes from original white to red or brown colour. This is called red or brown discolouration of white fish fillets. It occurs mainly because of two reasons. (1) If fish is filleted immediately after death without bleeding, blood remains in the flesh and clots there imparting brown colour to the originally white flesh. (2) If fish is subjected to rough handling while the fish is still alive, it may cause bruises, rupture of blood vessels and blood may ooze into the flesh (haematoma). However, heavy pressure on dead fish, when blood has already clotted (overloading of fish boxes) does not enhance discolouration. The brown discolouration of white fish fillets can be controlled in two ways. (1) Bleeding the fish immediately after harvesting for at least 15 minutes before being gutted. (2) Proper handling of the catch while the fish is still alive.

Glossary

Absolute sterility: It is the condition of a material attained due to absolute absence of microbes and any other forms of life in it by their complete destruction or eradication.

Acidification: It is the process by which a non-acid product is converted to acidic product by treatment with acid.

Aerobic glycolysis: It is the process by which each glucose molecule, in presence of oxygen (air), is converted to two pyruvic acid molecules in a sequence of steps, which in turn enter into the Krebs cycle. It is usually used synonymously to glycolysis.

Amino acid: Organic acid (*i.e.* acid containing a carboxylic, -COOH, group) with an amino ($-NH_2$) group.

Amylolytic path: It is a path of glycogenolysis (hydrolysis of glycogen) in which glycogen is directly hydrolysed to glucose by the enzyme α-amylase. The glucose produced is phosphorylated to produce glucose-6-phosphate, which goes for anaerobic glycolysis resulting in the production of lactic acid.

Anaerobic glycolysis: It is the process by which each glucose molecule, in absence of oxygen (air), is converted to two pyruvic acid molecules, which in turn are further converted to two lactic acid molecules in a sequence of steps.

Antibodies: Substances produced inside body of living fish in response to entry of foreign particles called antigens.

Antioxidants: These are the substances, which can act against lipid autoxidation.

Autocatalytic process: An autocatalytic process is one in which the end products of the reaction themselves catalyse the reaction.

Autolysis: It is the post mortem process in which, the complex tissue components of fish are hydrolysed or broken down into their simple building blocks by the endogenous enzymes present in the fish body.

Autophagy: It is a process by which the lysosomes of a cell secret their enzymes on intracellular organelles like mitochondria or golgi complex present within the same cell and digest them.

Autoxidation: It is the process by which unsaturated lipids of fish tissue are oxidised by the atmospheric oxygen at their points of unsaturation.

Available lysine: It is the amount of lysine present in a food or feed, which can be made available for utilisation by the consumer.

Belly bursting: It is a phenomenon occurring in some pelagic fishes in which the belly cavity starts swelling after death due to accumulation of gas produced inside it leading ultimately to its bursting.

Biogenic amines: These are the poisonous basic mono- and di-amines, which are formed through decarboxylation of amino acids and organic bases by the decarboxylase enzyme of putrefactive bacteria.

Carbohydrates: Carbohydrates are polyhydroxy aldehydes or ketones or substances that yield such compounds on hydrolysis.

Cataleptic rigor: When fish is killed by electric shock, rigor sets in long time after death. Such a rigor is called cataleptic rigor.

Chain initiators: These are the agents like light and metal ions, which initiate the formation of free radicals during the initiation step of lipid autoxidation.

Chain reaction: It is a series of reactions in which the product of the reaction of one molecule initiates the reaction of a second molecule, thus continuing the reaction in a chain.

Chilled seawater stowage (CSW): It is a short-term method of fish preservation in which the temperature of fish body is lowered by submerging it in seawater whose temperature has been reduced by addition of ice.

Cold storage dehydration: It is the loss of water from frozen fish during cold storage, which leads to their quality deterioration. It occurs when the relative humidity of the cold store atmosphere is less than that of the frozen fish.

Colour fixation: It is the treatment to fix the bright red colour of red meat of fishes like tuna so that it does not undergo discolouration to undesirable brown colour.

Commercial sterility: It is the condition of a fishery product attained by thermal processing such that the product, during its normal non-refrigerated condition of storage and distribution, neither spoils nor causes any health hazard in the consumer.

Curing: It is the process of mixing red meat of fishes like tuna with curing salts like sodium nitrite and keeping overnight so as to prevent the undesirable brown discolouration of the meat.

Deamination: It is the process of removal of the amino group ($-NH_2$) from an amino acid.

Decarboxylation: It is the process of removal of the carboxylic group (–COOH) from an amino acid with the production of CO_2 and an amine.

Decomposition: It is the third step in the autoxidation of lipids in which the feebly stable hydroperoxides, formed during propagation step, are decomposed into secondary oxidation products.

Discolouration: It is the undesirable changes in the original colour of fish tissue after death.

Drip loss: Drip loss is the loss of tissue fluid that drips out from frozen fishery products during their thawing.

Endocytosis: It is a process by which a cell engulfs a foreign particle into its cell after which its lysosomes release hydrolytic enzymes on the particle to digest it intracellularly.

Endogenous enzymes: These are the enzymes synthesised by the fish body and stored inside it for its own use in the metabolic processes as and when required.

Endogenous microbes: These are the microbes, which inhabit the body of fish when it is alive, but normally do not cause any disease in it.

Endotoxins: These are the integral part of the cell wall of infectious microbes consisting of lipid A fraction of the lipopolysaccharide (LPS) present in the outer membrane.

Essential fatty acids: The fatty acids, which cannot be synthesised in the body and must therefore be supplied in the diet are known as essential fatty acids. For most of the mammals linoleic and linolenic acids are the essential fatty acids.

Evisceration: Evisceration is the process of removal of the gut and internal organs from fish carcass.

Exocytosis: It is a process by which the lysosomes of a cell move to the cell membrane and rupture upon touching it releasing the lysosomal enzymes to outside where they digest any material extracellularly for further absorption into the cell.

Exogenous enzymes: These are the enzymes not synthesised by the fish body itself, rather those which catalyse reactions inside it when synthesised within its body by some other organisms like bacteria and protozoa present inside it or those which catalyse reactions when added to fish meat during post harvest processing.

Exogenous microbes: These are the microbes, which get their access to the fish body during post harvest handling, processing and storage.

Exotoxins: These are the proteinaceous substances synthesised within the cells of certain types of pathogenic bacteria and released to outside.

Extracellular enzymes: These are enzymes, which catalyse processes only outside the cells.

Fatty acid: Organic acid (*i.e.* acid containing a carboxylic, -COOH, group) with a hydrocarbon chain.

Fermentation: Fermentation is the process of break down of organic substances by enzymes under anaerobic condition.

Fillets: Fillets are the pieces of fish flesh obtained by cutting a fish longitudinally along both the sides of the backbone.

Fish chilling: Fish chilling is a short-term method of fish preservation in which the temperature of fish body is lowered to a temperature of about 0°C, which not only retards microbial activity considerably, but also slows down other spoilage processes remarkably.

Fish freezing: Fish freezing is a long-term method of fish preservation in which the temperature of fish body is lowered to a very low temperature, much below the freezing point of fish, by removal of heat from its body using artificial refrigeration system.

Fish irradiation: Fish irradiation is a method of fish preservation in which fish is exposed to an ionising radiation, which kills most of the bacteria present in it.

Fish pasteurisation: Fish pasteurisation is a short-term method of fish preservation in which fish or fishery products are exposed to moist heat reaching a maximum temperature of 100°C, the boiling point of water.

Fish spoilage: It is the condition of fish when the consumer rejects it due to its poor organoleptic qualities resulting from the cumulative accumulation of off-odour, off-flavour substances at very high concentrations.

Fish sterilisation: Fish sterilisation is a long-term method of fish preservation in which fish or fishery products are exposed to moist heat above 100°C, the boiling point of water, so that they become shelf-stable under normal non-refrigerated condition of storage and distribution.

Flavour reversion: It is the development of a particular type of oxidised off-flavour, described as "greasy", "beany", "fishy" or "painty" which occurs at comparatively low levels of oxidation of lipids containing the fatty acid linolenic acid upon exposure to air.

Food poisoning: It is the health hazard caused by the pathogenic microbes from consumption of fish and other foods.

Foreign enzymes: *See 'Exogenous enzymes'.*

Free fatty acids (FFA): It includes all the fatty acids present freely in fish tissue not bonded to glycerol molecules.

Free radical: A free radical is the component of a molecule, which has a free hand available due to removal of a hydrogen radical from it.

Functional properties of proteins: Functional properties of proteins are the physico-chemical properties, which influence the processing and behaviour of the proteins in food systems as judged by the quality attributes of the final product.

Gaping: Gaping is the phenomenon due to which individual flakes (myotomes) of fish muscle get apart giving the tissue a broken and ragged appearance.

Glaze water: It is the chilled potable water used for glazing of fish.

Glazing: It is the process adopted in freezing preservation of fish in which a thin layer of ice is allowed to be formed on the product so as to give it a protective coating.

Glycogenesis: It is the process by which the glucose molecules present in the fish tissue are polymerised to glycogen for storage.

Glycogenolysis: It is the process by which the glycogen stored in the fish body is hydrolysed to glucose for use.

Glycolysis: It is the process by which each glucose molecule, in presence of oxygen (air), is converted to two pyruvic acid molecules in a sequence of steps.

Glycolytic path: It is a path of glycogenolysis (hydrolysis of glycogen) in which one molecule of glucose from the chain of glucose molecules in glycogen is directly phosphorylated to glucose-1-phosphate by the enzyme phosphorylase a. Glucose-1-phosphate is isomerised to glucose-6-phoaphate, which goes for anaerobic glycolysis resulting in the production of lactic acid.

Harmful microbes: They are the microbes, which cause different types of harms to man.

Helpful microbes: They are the microbes, which help man in some way or the other.

Histamine poisoning: It is a type of poisoning caused by consumption of foods containing histamine in high concentration. It is also called scombroid poisoning as it very often occurs from the consumption of fishes belonging to the family Scombridae.

Histamine toxicity: *See 'Histamine poisoning'*.

Hydrogenation of fats: It is the process by which unsaturated fats are converted to saturated fats by addition of hydrogen atoms at the points of unsaturation with the help of catalysts.

Hydrolases: A class of enzymes, which catalyse hydrolysis reactions (transfer of functional groups to water).

Hydrolytic rancidity: It is the rancid flavour resulting from the accumulation of free fatty acids (FFA) produced by hydrolysis of tissue fats by lipolytic body enzymes (lipases) during autolysis or by lipolytic bacteria during microbial putrefaction.

Hyperaemia: The phenomenon of secretion of large quantity of slime to the body surface of fish after death is called hyperaemia.

Icing: It is a short-term method of fish preservation in which the temperature of fish body is lowered by keeping it in contact with ice.

Immunity: It is the defense mechanism of the body of living fish against invasion by microorganisms.

Immunoglobins: Protein particles produced in the body of living fish in response to entry of foreign particles into its body.

Induction period: It is the initial period of the propagation step of lipid autoxidation during which hydroperoxide formation takes place slowly. After this period, the hydroperoxide formation increases suddenly.

Infection-type food poisoning: It is the food poisoning caused by consumption of foods infected with certain types of living food poisoning microorganism, whose activity in the body of the consumer is essential for the food poisoning to occur.

Initiation: It is the first step in the autoxidation of lipids in which one hydrogen radical (H*) is removed from an unsaturated fatty acid of a lipid molecule (R_1H) at the carbon atom next to the double bond (*i.e.* the α-carbon to the double bond) resulting in a lipid free radical (R_1*).

Initiators: *See 'Chain initiators'.*

Intoxication-type food poisoning: It is the food poisoning caused by consumption of foods containing toxins produced by certain types of food poisoning microorganism.

Intracellular enzymes: These are enzymes, which catalyse processes only inside the cells.

K value: It is the percentage of inosine and hypoxanthine in total ATP and its degradation products present in fish tissue after death.

K_1 value: It is the percentage of inosine and hypoxanthine in total IMP, inosine and hypoxanthine present in fish tissue after death.

Lag of initiation: It is the time lag between death and onset of rigor mortis.

Lipid autoxidation: *See 'Autoxidation'.*

Lipid oxidation: *See 'Autoxidation'.*

Lipids: Lipids are a heterogeneous group of compounds, which are relatively insoluble in polar solvents like water and soluble in nonpolar solvents such as ether, chloroform and benzene. Thus, lipids include, fats, oils, waxes, fatty acids, fat-soluble vitamins etc.

Lipolysis: It is a process by which the tissue lipids of fish, mainly the fats, are broken down by a group of lipid-breaking or lipolytic (*lipo*: lipid; *lytic*: breaking) enzymes collectively called lipases.

Maillard reaction: It is the development of brown colour during the heating or prolonged storage of foods occurring non-enzymatically due to reactions between amines, amino acids or proteins with sugars, aldehydes or ketones.

Maturation: It is the process, taking place in salted fish during its prolonged storage at cooler temperature by which the endogenous enzymes hydrolyse the fish tissue components to simpler compounds responsible for the typical pleasant mature taste greatly relished by the consumers. It also results in the conversion of tough meat to tender and pliable meat.

Melanosis: It is a quality problem very often encountered in the post harvest storage of shrimps and other shellfishes by icing, in which there is formation of black spots on shells followed by a general blackening of the shell and meat.

Microbes: These are living forms whose size is so small that they are usually not visible to naked eye and can be seen only with the help of a microscope. They include bacteria, viruses, protozoa, microscopic fungi and algae.

Microbial decomposition: It is the post mortem process of conversion of fish tissue components by microorganisms into off-odour, off-flavour substances, accumulation of which distracts the consumers, and thereby results in the spoilage of fish.

Microbial putrefaction: *See 'Microbial decomposition'.*

Microorganisms: *See 'Microbes'.*

Modified atmosphere packaging: Modified atmosphere packaging (MAP) is a method adopted to extend the shelf life of fresh fish during chilled storage. In this method, fresh fish is packed in an airtight container within which the atmosphere is modified desirably by flushing-in of required gases (CO_2, N_2 and O_2) in required proportion.

Native enzymes: *See 'Endogenous enzymes'.*

Natural microbial flora: *See 'Endogenous microbes'.*

Non-enzymatic browning: *See 'Maillard reaction'.*

Non-protein nitrogen (NPN) content: It is the content of nitrogen in fish tissue contributed by all the nitrogenous substances except proteins.

Nucleolysis: It is a process by which the nucleic acids of fish tissues are hydrolysed by a group of endogenous enzymes collectively called nucleases.

Nucleotidolysis: Nucleotidolysis (*nucleotido*: of nucleotide; *lysis*: break down) or nucleotide degradation is the process of break down of nucleotides by a group of enzymes, collectively called nucleotidases.

On-rigor: It is the state of fish carcass from the onset of rigor mortis till its disappearance during which the fish body remains stiffened as rigor mortis continues.

Organoleptic evaluation: It is a method of evaluation of quality of fish and fishery products by human beings using their sense organs of vision, smell, taste, touch and hearing.

Original softness level: It is the level of softness of freshly killed fish.

Oxidative rancidity: It is the rancid flavour, which develops in fish after death or in fishery products during processing and storage due to accumulation of certain end products of lipid autoxidation.

Pathogenic microbes: These are the microbes whose presence or growth in fish and fishery products can cause health hazards to the person who consumes it.

Peroxide value (PV): It is a measure of the quantity of hydroperoxides produced during autoxidation of lipids, so as to indicate the extent of autoxidation and rancidity that the lipid has undergone. Thus, it is used as an indicator of the extent of lipid autoxidation.

Phosphocreatine: It is an energy-rich phosphate compound present in fish tissue, which dissociates to creatine and phosphate with liberation of energy. This energy is used to resynthesise ATP from ADP and Pi so as to replenish the ATP utilised in various metabolic processes and maintain it at a constant level.

Polymerisation: It is a type of lipid deterioration, which occurs only when lipids are exposed to high temperature processing such as canning and frying. In the process of polymerisation, single units of fatty acid monomers combine in a chain manner at their respective points of unsaturation to form a chain or polymer of fatty acids.

Post mortem changes: The changes that take place in fish body after death are collectively called post mortem changes. They include hyperaemia, rigor mortis, autolysis, microbial putrefaction, lipid autoxidation and discolouration.

Post-freezing glaze treatment: It is the process adopted after freezing of fishes in which a thin layer of ice is allowed to be formed on the frozen product, either by spraying chilled water to its surface or by giving it a quick dip treatment in chilled water so as to give it a protective coating.

Post-rigor: It is the state of fish carcass after the disappearance of rigor mortis, during which the muscle becomes soft and flabby.

Pre-freezing glaze treatment: It is the process adopted before freezing of fishes in which chilled water is poured into fishes arranged in a container (usually a rectangular tray) so that the water fills the void space between fishes and completely covers the product from all sides thereby giving a uniform block after freezing.

Pre-rigor: It is the state of fish carcass from death till the onset of rigor mortis during which the fish is extremely fresh and the muscle remains elastic.

Primary oxidation products: These are the first group of stable products, such as hydroperoxides, formed in the propagation step of lipid autoxidation.

Proenzymes: These are enzymes in their inactive precursor form present within the secretory cells, which are converted to the active enzyme form once secreted to outside the cells.

Prooxidants: These are the substances, which accelerate the rate of lipid autoxidation.

Propagation: It is the second step in the autoxidation of lipids in which the few free radicals formed in the initiation step propagate to produce large number of free radicals.

Protein denaturation: Any change in the native three-dimensional conformation of proteins is called protein denaturation.

Proteins: Proteins are chains of covalently linked amino acid residues, which spontaneously acquire three-dimensional conformations based on the self-recognition of the amino acid sequences.

Proteolysis: It is a process by which the tissue proteins of fish are broken down by a group of endogenous protein-breaking or proteolytic (*proteo*: proteins; *lytic*: breaking) enzymes collectively called proteases.

Ptomaines: *See 'Biogenic amines'*.

Putrefactive microbes: These are the microbes causing decomposition of the fish tissue components to off-odour, off-flavour substances, accumulation of which distracts the consumer, and thereby renders the fish spoiled. However, if consumed, it may not lead to any health hazard.

Putrescent alkaloids: *'See Biogenic amines'*.

Rancid flavour: It is the typical off-flavour, which develops in lipid-containing foods due to the deterioration of the lipids present in them.

Rancidity: *See 'Rancid flavour'*.

Refining of oil: It is the process by which the free fatty acids (FFA) present in the oil are removed by forming soaps with alkali.

Refrigerated seawater storage (RSW): It is a short-term method of fish preservation in which the temperature of fish body is lowered by submerging it in seawater whose temperature has been [illegible]frigeration system.

[illegible]robes: *See 'Endogenous microbes'*.

Resonance hybrid: A pair of forms of a molecule resulting from resonance stabilisation.

Resonance stabilisation: It is the process by which a free radical is stabilised by the movement of the free hand from one atom to another.

Rigor mortis: It is a post mortem change wherein the fish body, after a short lapse of time from death, remains stiffened for a certain period during which the elastic nature of living muscle disappears and the muscle becomes hard.

Ripening: *See 'Maturation'*.

Saccharolysis: Saccharolysis (*saccharo*: sugar; *lysis*: break down) is the process of breaking down the carbohydrates or sugar substances by a group of carbohydrate-breaking or saccharolytic enzymes, collectively called carbohydrases.

Scombroid poisoning: *See 'Histamine poisoning'*.

Secondary oxidation products: These are the second group of stable products, such as aldehydes, alcohols and ketones formed in the decomposition step of lipid autoxidation.

Sensory evaluation: *See 'Organoleptic evaluation'.*

Sliding filament-cross bridge theory: According to this theory, due to the disappearance of ATP, the actin filaments form cross bridges (bonds) with myosin filaments after which the actin filaments slide past the myosin filaments towards the center of the sarcomere. This results in the contraction of the muscle fibers, which leads to the stiffening of muscle during rigor mortis.

Spoilage microbes: *See 'Putrefactive microbes'.*

Synergists: These are the substances, which are not highly effective by themselves as antioxidants against lipid autoxidation, but when used together with antioxidants, increase the effectiveness of the antioxidants.

Tenderisation: It is the process of conversion of tough meat of fresh fish to tender, pliable organoleptically better meat by the autolytic hydrolysis of the fish proteins by endogenous enzymes.

Termination: It is the last step of lipid autoxidation in which the chain reaction of autoxidation process is terminated as the free radicals react within themselves to yield non-reactive products.

Tertiary oxidation products: These are the third group of stable products, such as short chain fatty acids formed in the decomposition step of lipid autoxidation.

Thaw rigor: When fish is frozen without completing the rigor mortis (pre-rigor or on-rigor), the freezing process arrests the rigor process and the rigor is completed during thawing. This rigor is called thaw rigor.

Thiobarbituric acid number (TBA Number): It is a measure of the quantity of malonaldehyde produced during autoxidation of lipids, which indicates the extent of autoxidation that the lipid has undergone. Thus, it is used as an indicator of the extent of lipid autoxidation.

Water activity (a_w): It is the ratio of vapour pressure of fish tissue to that of pure water. It is used to indicate the level of moisture available in fish tissue for microbial and chemical activities.

Zymogens: *See 'Proenzymes'.*

Bibliography

Almas, K. A., 1981. *Chemistry and Microbiology of Fish and Fish Processing*, Department of Biochemistry, Norwegian Institute of Technology, University of Trondheim, Trondheim, Norway, 123 pp.

American Meat Institute Foundation, 1960. *The Science of Meat and Meat Products*, W.H. Freeman and Company, San Francisco, 438 pp.

Amlacher, E., 1961. Rigor mortis in fish. *In*: *Fish as Food*, Volume I (Borgstrom, G., Ed.), Academic Press Inc., New York, pp. 385-409.

Balachandran, K. K., 2001. *Post-Harvest Technology of Fish and Fish Products*, Daya Publishing House, New Delhi, 440 pp.

Banwart, G. J., 1974. *Basic Food Microbiology*, CBS Publishers and Distributors, New Delhi, 781 pp.

Cappuccino, J. G. and Sherman, N., 1987. *Microbiology- A Laboratory Manual*, The Benjamin/Cummings Publishing Company, Inc., California, 458 pp.

CIFT, 1991. *Manual of Summer Institute on Modern Techniques of Testing Fish and Fishery Products for Food and Feed*, 6th to 25th May, 1991, Biochemistry, Nutrition and Microbiology Division, Central Institute of Fisheries Technology (CIFT), Matsyapuri , Cochin.

CIFT, 2003. *Product Development and Seafood Safety*, (Joseph, J., Mathew, P.T., Joseph, A.C. and Muraleedharan, C., Eds.), Central Institute of Fisheries Technology (CIFT), Matsyapuri, Cochin, 467 pp.

Coleby, B., 1965. The radiation preservation of fish, *In*: *Fish as Food*, Volume III (Borgstrom, G., Ed.), Academic Press Inc., New York, pp. 419-435.

Conn, E. E. and Stumph, P. K., 1976. *Outlines of Biochemistry*, Wiley Eastern Limited, New Delhi, 629 pp.

Connell, J. J., 1980. *Advances in Fish Science and Technology*, Fishing News Books Ltd., Farnham, Surrey, England, 512 pp.

De Man, J. M., 1976. *Principles of Food Chemistry*, The AVI Publishing Company, Inc., Westport, Connecticut, 426 pp.

De Robertis, E. D. P. and De Robertis, Jr., F. M. F., 1980. *Cell and Molecular Biology*, Saunders College, Philadelphia, 673 pp.

Frazier, W. C., 1958. *Food Microbiology*, McGraw-Hill Book Company, Inc., New York, 472 pp.

Giese, A. C., 1979. *Cell Physiology*, W.B. Saunders Company, Philadelphia, 609 pp.

Gopakumar, K., 2002. *Text Book of Fish Processing Technology*, Indian Council of Agricultural Research, New Delhi, 491 pp.

Hoar, W. S., 1987. *General and Comparative Physiology*, Prentice-Hall of India Private Limited, New Delhi, 851 pp.

Huss, H. H., 1994. *Assurance of Seafood Quality*, FAO Fisheries Technical Paper, No 334, FAO, Rome, 169 pp.

Huss, H. H., 1995. *Quality and Quality Changes in Fresh Fish,* FAO Fisheries Technical Paper, No 348, FAO, Rome, 195 pp.

Lawrie, R. A., 1974. *Meat Science*, Pergamon Press, New York, 419 pp.

Lehninger, A. L., 1984. *Principles of Biochemistry*, CBS Publishers and Distributors, New Delhi, 1011 pp.

Madigan, M. T., Martinko, J. M. and Parker, J., 2000. *Brock Biology of Microorganism,* Prentice-Hall, Inc., New Jersey, 991 pp.

Murray, R. K., Granner D. K., Mayes, P. A. and Rodwell, V. W., 1993. *Harper's Biochemistry*, Prentice-Hall of India Private Limited, New Delhi, 806 pp.

Salle, A. J., 1988. *Fundamental Principles of Bacteriology,* Tata McGraw-Hill Publishing Company Ltd., New Delhi, 1094 pp.

Sen, D. P., 2005. *Advances in Fish Processing Technology*, Allied Publishers Pvt. Ltd., New Delhi, 818 pp.

Stroud, G. D., *Rigor Mortis in Fish: The Effect on Quality*, Torry Advisory Note No. 36. Torry Research Station, Aberdeen, UK.

Stryer, L., 1998. *Biochemistry*, W. H. Freeman and Company, New York, 1064 pp.

Suzuki, T., 1981. *Fish and Krill Proteins: Processing and Technology,* Applied Science Publishers Ltd., London, England, 260 pp.

Tsuchiya, T., 1961. Biochemistry of fish oils. *In* : *Fish as Food*, Volume I (Borgstrom, G., Ed.), Academic Press Inc., New York, pp. 211-274.

Wheaton, F. W. and Lawson, T. B., 1985. *Processing Aquatic Food Products*. John Wiley & Sons, New York, 492 pp.

Zaitsev, V., Kizevetter, I., Lagunov, L., Makarova, T., Minder, L. and Podsevalov, V., 1969. *Fish Curing and Processing*, MIR Publishers, Moscow, 721 pp.

Index